Essential SQA Exam Practice

Higher Human Biology Practice Questions & Exam Papers

Questions & Papers

- Practise **100+ questions** covering every question type and topic
- Complete **2 practice papers** that mirror the real SQA exams

Billy Dickson
Graham Moffat

HODDER GIBSON
AN HACHETTE UK COMPANY

Although every effort has been made to ensure that website addresses are correct at time of going to press, Hodder Gibson cannot be held responsible for the content of any website mentioned in this book. It is sometimes possible to find a relocated web page by typing in the address of the home page for a website in the URL window of your browser.

Hachette UK's policy is to use papers that are natural, renewable and recyclable products and made from wood grown in well-managed forests and other controlled sources. The logging and manufacturing processes are expected to conform to the environmental regulations of the country of origin.

Orders: please contact Bookpoint Ltd, 130 Park Drive, Milton Park, Abingdon, Oxon OX14 4SE. Telephone: (44) 01235 827827.
Fax: (44) 01235 400401. Email education@bookpoint.co.uk. Lines are open from 9 a.m. to 5 p.m., Monday to Friday, with a 24-hour message answering service. Visit our website at www.hoddereducation.co.uk. If you have queries or questions that aren't about an order you can contact us at hoddergibson@hodder.co.uk

© Billy Dickson and Graham Moffat 2019

First published in 2019 by
Hodder Gibson, an imprint of Hodder Education
An Hachette UK Company
211 St Vincent Street
Glasgow, G2 5QY

Impression number	5	4	3	2	1
Year	2023	2022	2021	2020	2019

All rights reserved. Apart from any use permitted under UK copyright law, no part of this publication may be reproduced or transmitted in any form or by any means, electronic or mechanical, including photocopying and recording, or held within any information storage and retrieval system, without permission in writing from the publisher or under licence from the Copyright Licensing Agency Limited. Further details of such licences (for reprographic reproduction) may be obtained from the Copyright Licensing Agency Limited, www.cla.co.uk.

Illustrations by Aptara Inc.
Typeset in India by Aptara Inc.
Printed and bound by CPI Group (UK) Ltd, Croydon CR0 4YY
A catalogue record for this title is available from the British Library.
ISBN: 978 1 5104 7180 1

CONTENTS

Introduction	iv
Key Area index grids	viii
Practice Questions	**1**
Question type: Multiple-choice	1
Question type: Structured and extended response	20
Short-answer questions	20
Data-handling questions	30
Experimental questions	35
Mini extended response questions	39
Full extended response questions	40
Answers to Practice Questions	**41**
Practice Exams	**52**
Practice Exam A	52
Practice Exam B	75
Answers to Practice Exams	**101**
Answers to Practice Exam A	101
Answers to Practice Exam B	112
Revision Calendar	**123**

INTRODUCTION

Higher Human Biology

The assessment materials included in this book are designed to provide practice and to support revision for the Higher Human Biology course examination, which is worth 80% of the final grade for this course.

The materials are provided in two sections:

1 Practice Questions
2 Practice Exams: Practice Exam A and Practice Exam B

Together, these give overall and comprehensive coverage of the assessment of

- **Knowledge and its Application**: demonstrating knowledge and understanding (KU) and applying knowledge and understanding
- **Skills of Scientific Inquiry** (SSI): planning, selecting, presenting, processing, predicting, concluding, evaluating.

(See the section on student margins on page v for the specific requirements of these types of assessment.)

The Higher Human Biology course is split into three areas of study: 1 Human cells; 2 Physiology and Health; 3 Neurobiology and immunology. These areas are further divided into **Key Areas**, which we have numbered 1.1, 1.2, ... , 2.1, 2.2, ... , 3.1, 3.2, ... , etc. Please refer to the course specification document, which can be found at www.sqa.org.uk/sqa/47915.html, for full descriptions of the requirements of each Key Area. This is an important tool in your revision.

Practice Questions

The practice questions have been written to be similar to those you will find in a Higher Human Biology examination and are arranged in sets to reflect the different types of question that are used there, as shown in the grid below.

There are **300 marks'** worth of questions in this section.

1 – Multiple choice	2 – Structured and extended response questions				
	Short answer	Data handling	Experimental	Mini extended response	Full extended response
Objective items testing knowledge and understanding (KU) and some testing SSI	Structured questions testing KU within Key Areas	Structured questions testing data-handling skills	Structured questions testing experimental skills	Extended writing questions testing related KU from a Key Area Usually worth 4–6 marks	Extended writing questions testing related KU from a Key Area Usually worth 7–10 marks
		(SSI: planning, selecting, presenting, processing, predicting, concluding, evaluating)			

Practice Questions Key Area index

The Practice Question index on page viii shows the pattern of coverage of the knowledge in the Key Areas of the specification, across the practice questions.

After working on questions from Key Areas across an area of study, you might want to use the boxes to assess your progress. We suggest marking like this [–] if you are having difficulty (less than half marks), like this [+] if you have done further work and are more comfortable (more than half marks) and this [*] if you are confident you have learned and understood an entire area of study (nearly full marks). Alternatively, you could use a 'traffic light' system using colours – red for 'not understood'; orange for 'more work needed' and green for 'fully understood'. **If you continue to struggle with a set of Key Area questions, you should see your teacher for extra help**.

›› HOW TO ANSWER

Each set of questions starts with a short commentary which describes the question type and gives some support in answering them.

INTRODUCTION

Practice Exams

Each paper has been carefully written to be very similar to a typical Higher examination question paper. Your Higher examination has 120 marks in total and is divided into two papers:

Paper 1 – Objective test, which contains 25 multiple-choice items worth 1 mark each and totalling 25 marks altogether.

Paper 2 – Structured questions, which contains restricted and extended response questions totalling 95 marks.

In each Practice Exam, the marks are distributed evenly across all three component areas of study of the course and in each paper about 90 marks are for the demonstration and application of Knowledge and Understanding. The remaining marks are for the application of Skills of Scientific Inquiry.

70% of the marks in each Practice Exam are set at the demand level of Grade C and the remaining 30% are more challenging marks set at the demand level for Grade A. We have attempted to construct each Practice Exam to represent the typical range of demand in a Higher Human Biology paper.

Grading

The two Practice Exams are designed to be equally demanding and to reflect the National Standard of a typical SQA exam. Each exam has 120 marks – if you score 60 marks that's a C pass. You will need about 72 marks for a B pass and about 84 marks for an A. These figures are a rough guide only.

Timing

If you are attempting a full Practice Exam, limit yourself to **3 hours** to complete it. Get someone to time you! You should take no more than 40 minutes for Paper 1 and no more than 2 hours and 20 minutes for Paper 2.

If you are tackling blocks of questions in a Key Area or by question type, give yourself about a minute and a half per mark. For example, a set of questions worth 10 marks should take about 15 minutes.

Practice Exam index

The Practice Exam index on pages ix and x shows the pattern of coverage of the knowledge in the Key Areas of the specification, across the two papers.

We have provided a marks total for each area of study and each question type. You could use the check boxes to record your total mark for each area of study. Scoring more than half of these marks suggests you have a good grasp of the content of that area of study.

Student margins

All questions in both the Practice Questions and Practice Exams have margins. The margins have a key to each question to show what is being tested, as shown in the table below.

	Student margin key	Meaning
Knowledge and its Application	Demonstrating KU	Demonstrating knowledge and understanding of human biology
	Applying KU	Applying knowledge of human biology to new situations, analysing information and solving problems
Skills of Scientific Inquiry (SSI)	Planning	Planning or designing experiments/practical investigations to test given hypotheses
	Selecting	Selecting information from a variety of sources
	Presenting	Presenting information appropriately in a variety of forms
	Processing	Processing information (using calculations and units, where appropriate)
	Predicting	Making predictions and generalisations based on evidence/information
	Concluding	Drawing valid conclusions and giving explanations supported by evidence/justification
	Evaluating	Suggesting improvements to experiments and practical investigations

INTRODUCTION

Techniques

There are **six laboratory techniques** with which you should be familiar for your Higher Human Biology exam. You might even use one of these for your assignment. The grid below shows the questions throughout this book that are related to these techniques. If you are not sure about any of the techniques or the questions referenced, you should ask your teacher for some advice.

Technique	Practice Questions					Practice Exams			
						Exam A		Exam B	
	Multiple choice	Structured and extended response				Paper 1	Paper 2	Paper 1	Paper 2
		Short answer	Data	Experimental	Extended response	Multiple choice	Structured and extended response	Multiple choice	Structured and extended response
1 Using gel electrophoresis	4			1					
2 Altering enzyme reaction rates	51					6			4
3 Measuring metabolic rate			1						
4 Using a respirometer	54								
5 Measuring pulse rate and blood pressure				2		14			11
6 Measuring body mass index (BMI)	56		2						

Using the questions and practice exams

We recommend working between attempting questions and studying the answers (see below).

Each Practice Exam can be attempted in one session, or groups of questions on a particular Key Area or question type can be tackled – use the question indexes to find related groups of questions.

Where any difficulty is encountered, it's worth trying to consolidate your knowledge and skills. Use the information in the student margin to identify the type of question you find trickiest. Be aware that Grade A-type questions are expected to be challenging.

You will need a pen, a sharp pencil, a clear plastic ruler and a calculator for the best results. A couple of different coloured highlighters could also be handy.

Answers

The expected answers for the Practice Questions are provided on pages 41–51 and those for the Practice Exams on pages 101–122. They give National Standard answers but, occasionally, there may be other acceptable answers.

Each answer has a reference to the level of demand of the question – C for questions at a demand level of a Grade C, and A for those with a demand level of a Grade A. For questions worth more than 1 mark, we have allocated each mark to a demand level. For example, CA means 1 mark at C and the second at A.

The answers to the Practice Exams also have commentaries with hints and tips provided alongside. Don't feel you need to use them all!

The commentaries on the answers focus on the biology itself, as well as hints and tips, advice on wording of answers and notes of commonly made errors.

Revision

There are 24 Key Areas, so covering two each week would need a 12-week revision programme. Starting during your February holiday should give you time for the exam in May – just! You could use the Revision Calendar on page 123 or at www.hoddergibson.co.uk/ESEP-extras to help you plan and keep a record of your progress.

We wish you the very best of luck!

KEY AREA INDEX GRIDS

Practice Questions

This Key Area index grid will guide you when looking for questions by question type or by Area of Study.

Course Areas		Multiple choice (1 mark)	Structured and extended response (ER) questions					Check
Area of Study	Key Area		Short answer (4 marks)	Data handling (8 marks)	Experimental (8 marks)	Mini ER (3–5 marks)	Full ER (9 marks)	
1 Human cells	1.1	1–2	1	1	1	1	1	/96
	1.2	3–4	2					
	1.3	5–6	3			2		
	1.4	7–8	4					
	1.5	9–10	5				2	
	1.6	11–12	6					
	1.7	13–14	7			3		
	1.8	15–16	8					
2 Physiology and health	2.1	17–18	9	2	2	4	3	/96
	2.2	19–20	10					
	2.3	21–22	11					
	2.4	23–24	12					
	2.5	25–26	13					
	2.6	27–28	14			5	4	
	2.7	29–30	15			6		
	2.8	31–32	16					
3 Neurobiology and immunology	3.1	33–34	17	3	3	7	5	/96
	3.2	35–36	18					
	3.3	37–38	19					
	3.4	39–40	20			8		
	3.5	41–42	21					
	3.6	43–44	22				6	
	3.7	45–46	23			9		
	3.8	47–48	24					
SSI		49–60						/12
Totals		60	96	24	24	42	54	300

Practice Exam A

This Key Area index grid will guide you when looking for questions by question type or by Area of Study.

Course Areas		Paper 1	Paper 2					Check
Area of Study	Key Area	Multiple choice	Short answer	SSI Data handling	SSI Experimental	Mini extended response	Full extended response	
1 Human cells	1.1	3	1	5				38
	1.2	1, 2, 5						
	1.3	4	2					
	1.4		3					
	1.5							
	1.6	6, 7, 8, 9	4					
	1.7		6					
	1.8							
2 Physiology and health	2.1	10				13		40
	2.2		7					
	2.3	11, 17						
	2.4	18	8					
	2.5	12, 13						
	2.6	14	11, 12					
	2.7	15	10					
	2.8		9					
3 Neurology and immunology	3.1	19			15		19	42
	3.2		16					
	3.3	21						
	3.4	20	14					
	3.5							
	3.6	22, 24	18					
	3.7	16, 23, 25						
	3.8		17					
Totals		25	67	8	8	4	8	120

Practice Exam B

This Key Area index grid will guide you when looking for questions by question type or by Area of Study.

Course Areas		Paper 1	Paper 2					Check
Area of Study	Key Area	Multiple choice	Short answer	SSI Data handling	SSI Experimental	Mini extended response	Full extended response	
1 Human cells	1.1	1, 2		4	3			$\overline{37}$
	1.2		1, 2					
	1.3	3, 4						
	1.4	6						
	1.5	7						
	1.6	5						
	1.7		5					
	1.8	8, 9						
2 Physiology and health	2.1		6	11			17	$\overline{37}$
	2.2							
	2.3							
	2.4	15, 16						
	2.5							
	2.6	10, 11	13					
	2.7	12						
	2.8	13, 14	7					
3 Neurology and immunology	3.1	21	9					$\overline{46}$
	3.2		8					
	3.3	18, 22						
	3.4	17	12, 14					
	3.5							
	3.6	20, 23, 24	15					
	3.7	19	10					
	3.8	25	16					
Totals		25	66	6	9	4	10	120

PRACTICE QUESTIONS

Question type: Multiple-choice

≫ HOW TO ANSWER

In your examination, Paper 1 consists entirely of multiple-choice questions. There are 25 questions for 1 mark each. Each question should take about 1.5 minutes and has only **one** correct answer.

In practice, some questions might take a bit longer, for example if there is a lot to read or if calculations or other information processing are involved. Others can be answered more quickly if they involve straightforward recall. The time for these questions is taken up in reading and thinking – there is no writing, only a mark in a grid, although you may need to do some rough working.

When tackling multiple-choice questions, read the question thoroughly and try to think of the answer without studying the options. Then look at the options:

▶ If your answer is there, that's the job done.
▶ If you are not certain of an answer, read through the question again and choose the option that seems the best fit.
▶ Or, you can try to eliminate options that you are sure are not correct, before making your choice.

> **Top Tip!**
> You should spend no more than 40 minutes on Paper 1 in your examination.

Try not to leave any question without an answer marked – complete the grid for each question as you work through.

For these multiple-choice practice questions, you may circle the letter corresponding to your chosen answer, or write your answers on a separate piece of paper.

> **Top Tip!**
> In your examination, any rough working for Paper 1 should be done on the additional space for answers and rough work, provided at the end of the supplied answer booklet.

---

1 There are two types of human stem cell.
   1 embryonic stem cells
   2 tissue (adult) stem cells

Which row in the table identifies the properties of embryonic and tissue stem cells?

	Properties of stem cell		
	Self-renewal	Can differentiate	Are multipotent
A	1 only	1 only	both 1 and 2
B	both 1 and 2	both 1 and 2	2 only
C	1 only	both 1 and 2	1 only
D	both 1 and 2	1 only	both 1 and 2

2 Cellular differentiation occurs because
   A cells express some of their genes but not others
   B cells all have a different genetic composition
   C different cells contain a different set of chromosomes
   D different cells lack some genes.

STUDENT MARGIN

Demonstrating KU

Demonstrating KU

1

PRACTICE QUESTIONS

3 The diagram represents the components of a single DNA nucleotide.

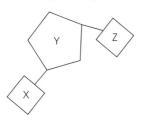

Which row in the table identifies the components in the diagram?

	X	Y	Z
A	phosphate	sugar	base
B	base	sugar	phosphate
C	sugar	phosphate	base
D	sugar	base	phosphate

Applying KU

4 The diagram shows a technique that can be used to separate fragments of DNA to produce a profile.

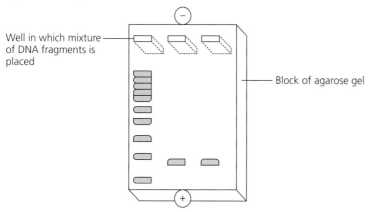

This technique is

A electrophoresis, where fragments are separated according to their size and charge

B PCR, where fragments are separated according to their size and charge

C electrophoresis, where fragments are separated according to their solubility in agarose gel

D PCR, where fragments are separated according to their solubility in agarose gel.

Applying KU

5 The list shows steps in the synthesis of the protein insulin.

1 transcription of DNA
2 polypeptide chains fold
3 RNA splicing
4 translation of mRNA

Which of the following sequences shows the order in which these steps occur?

A 1, 3, 4, 2
B 1, 4, 2, 3
C 3, 1, 2, 4
D 3, 4, 2, 1

Demonstrating KU

6 The diagram represents the chemical structure of a folded polypeptide.

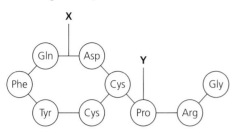

Which row in the table identifies parts X and Y in the diagram?

	X	Y
A	peptide bond	base
B	peptide bond	amino acid
C	hydrogen bond	base
D	hydrogen bond	amino acid

7 The list shows different mutations that can affect a single gene.
1   nonsense
2   missense
3   frameshift

Which of these gene mutations always results in the synthesis of a polypeptide that contains significantly fewer amino acids than would normally be expected?

A   1 only
B   1 and 2 only
C   1 and 3 only
D   3 only

8 Chronic myeloid leukaemia (CML) is a type of cancer that affects the white blood cells. Most people with CML have an abnormal chromosome called the Philadelphia chromosome. This happens when the *ABL1* gene on chromosome 9 breaks off and sticks to the *BCR* gene on chromosome 22.

From this information, it is possible to state that this genetic abnormality is an example of

A   translocation
B   substitution
C   duplication
D   insertion.

9 Which of the following is an application of pharmacogenetics?

A   comparing genomic sequences to help solve crimes
B   analysing genomic sequences to diagnose genetic disorders
C   predicting the likelihood of developing certain diseases
D   using genomic sequencing to select effective drug treatments

10 Bioinformatics is the

A   use of genome information in prescribing drugs
B   study of evolutionary history and relationships
C   use of computers and statistical analysis to compare sequence data
D   construction of molecular clocks to study evolutionary events.

PRACTICE QUESTIONS

11 The graph shows the energy changes involved in a chemical reaction.

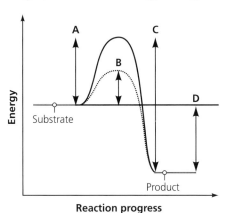

Which letter on the graph indicates the activation energy of this reaction in the presence of an enzyme specific to this substrate?

Applying KU

12 The diagram shows substances in a branched metabolic pathway.
Substance 1 is in steady supply from the diet.

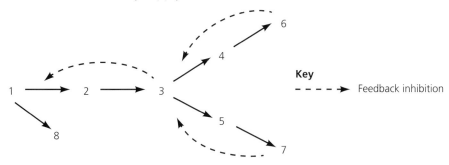

Which reaction would speed up if substances 6 and 7 were present in high concentrations?

A 3 → 4
B 2 → 3
C 5 → 7
D 1 → 8

Applying KU

13 The list shows types of reaction involved in respiration.

1 ATP synthesis
2 dehydrogenation
3 fermentation

Which row in the table identifies the types of reaction occurring at the given stages in respiration?

	Stages in respiration	
	Glucose converted to pyruvate in glycolysis	Citrate converted to oxaloacetate in the citric acid cycle
A	1 and 2	2 and 3
B	1 only	2 only
C	1 and 2	1 and 2
D	3 only	1 and 2

Demonstrating KU

PRACTICE QUESTIONS

14 The list shows processes in respiration.
   X   glycolysis
   Y   citric acid cycle
   Z   fermentation in muscle cells
   Which of these processes produces CO_2?
   A   X only
   B   Y only
   C   X and Y only
   D   X, Y and Z

15 Which pathway could occur in a muscle cell in the absence of sufficient oxygen?
   A   lactate → pyruvate → water and carbon dioxide
   B   glucose → pyruvate → acetyl coenzyme A
   C   glucose → lactate → water and carbon dioxide
   D   glucose → pyruvate → lactate

16 Which row in the table shows the properties of slow-twitch muscle fibres?

	Relative number of mitochondria	Concentration of myoglobin	Major storage fuel
A	many	low	fats
B	many	high	fats
C	few	low	glycogen
D	few	high	glycogen

17 The diagram shows a section through part of the testes.
   Which letter identifies cells that divide by both mitosis and meiosis?

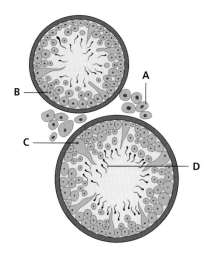

18 One function of the seminal vesicles is to
   A   store sperm
   B   produce ICSH
   C   maintain viability of sperm
   D   produce sperm.

PRACTICE QUESTIONS

19 The diagram represents the days of a normal 30-day menstrual cycle.

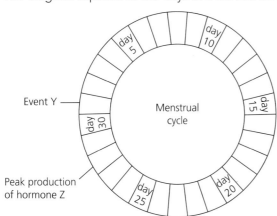

Which row in the table identifies event Y and hormone Z?

	Event Y	Hormone Z
A	ovulation	progesterone
B	onset of menstruation	oestrogen
C	ovulation	oestrogen
D	onset of menstruation	progesterone

Applying KU

20 The diagram shows the roles of three hormones in the feedback control of sperm production.

Key

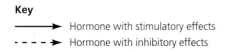

⟶ Hormone with stimulatory effects
- - -▶ Hormone with inhibitory effects

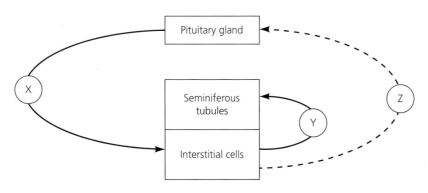

Which row in the table identifies the hormones involved at positions X, Y and Z in the diagram?

	X	Y	Z
A	ICSH	testosterone	ICSH
B	ICSH	testosterone	testosterone
C	FSH	ICSH	testosterone
D	FSH	ICSH	ICSH

Applying KU

PRACTICE QUESTIONS

21. The following procedures can be used in infertility treatments.
    1. artificial insemination (AI)
    2. intracytoplasmic sperm injection (ICSI)
    3. pre-implantation genetic diagnosis (PGD)

    Which of these procedures require *in vitro* fertilisation (IVF) as part of the fertility treatment?
    - A 1 and 2 only
    - B 2 and 3 only
    - C 1 and 3 only
    - D 3 only

22. *In vitro* fertilisation (IVF) is used to treat infertility.
    IVF involves using
    - A FSH to stimulate the maturation of eggs and LH to stimulate egg release
    - B LH to stimulate the maturation of eggs and FSH to stimulate egg release
    - C FSH to stimulate the maturation of eggs and progesterone to stimulate egg release
    - D LH to stimulate the maturation of eggs and progesterone to stimulate egg release.

23. The family history chart shows the inheritance of the metabolic disorder cystic fibrosis (CF) through three generations of a family. The allele coding for CF is autosomal recessive.

    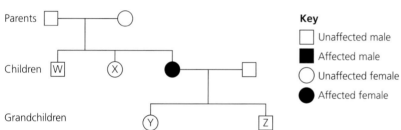

    Which of the individuals in the chart **must** be heterozygous for the CF allele?
    - A W, X, Y and Z
    - B W and X only
    - C Y and Z only
    - D X and Y only

24. Phenylketonuria (PKU) is a condition that results from
    - A differential gene expression
    - B a chromosome mutation
    - C an autosomal dominant disorder
    - D a substitution mutation.

PRACTICE QUESTIONS

25 The diagram represents a section through a blood vessel that carries blood away from the heart.

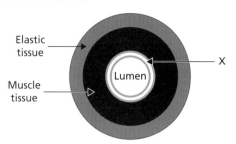

Which row in the table identifies the type of blood vessel and region X correctly?

	Type of blood vessel	Region X
A	artery	connective tissue
B	artery	endothelium
C	vein	connective tissue
D	vein	endothelium

Applying KU

26 The diagram represents part of a capillary bed involved in the exchange of materials with body cells.

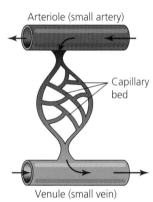

Which row in the table identifies the changes in the concentration of glucose, oxygen and carbon dioxide in the plasma as the blood flows through the capillary?

	Glucose	Oxygen	Carbon dioxide
A	increases	decreases	increases
B	decreases	increases	decreases
C	increases	increases	decreases
D	decreases	decreases	increases

Applying KU

27 The diagram shows a section through the heart.
At which point on the diagram would the atrioventricular node (AVN) be found?

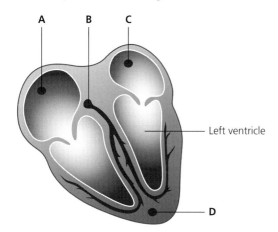

28 Which row in the table describes the state of the heart valves during atrial and ventricular diastole?

	Atrioventricular valves	Semi-lunar valves
A	open	open
B	closed	open
C	open	closed
D	closed	closed

29 The list shows events involved in the formation of a thrombus.
1. Clotting factors are released.
2. Insoluble fibres form a meshwork.
3. Soluble fibrinogen is converted to fibrin.
4. Inactive prothrombin is converted to its active form.

Which of the following sequences shows the order in which these events occur?
- A 1, 4, 3, 2
- B 1, 3, 4, 2
- C 4, 2, 3, 1
- D 4, 3, 1, 2

30 The ratio of high- to low-density lipoproteins in the blood (HDL:LDL) is related to the level of cholesterol in the blood and the chances of developing atherosclerosis.
Which row in the table correctly illustrates these relationships?

	HDL:LDL	Cholesterol level in the blood	Chance of developing atherosclerosis
A	higher	lower	reduced
B	higher	higher	increased
C	lower	lower	increased
D	lower	higher	reduced

PRACTICE QUESTIONS

31 The diagram shows some stages in the control of blood glucose concentration.

Stage 1 Increase in blood glucose concentration above normal is detected by receptors in the pancreas
↓
Stage 2 Increased secretion of hormone X; decreased secretion of hormone Y
↓
Stage 3 Organ Z converts glucose to storage carbohydrate
↓
Stage 4 Blood glucose concentration returns to normal
↓
Stage 5 Corrective mechanism switched off

Which row in the table identifies hormones X and Y and organ Z?

	Hormone X	Hormone Y	Organ Z
A	insulin	glucagon	pancreas
B	glucagon	insulin	liver
C	glucagon	insulin	pancreas
D	insulin	glucagon	liver

32 Which of the following describes typical features of type 1 diabetes?

	Feature of type 1 diabetes	
A	occurs in childhood	cells less sensitive to insulin
B	develops later in life	cells unable to produce insulin
C	occurs in childhood	cells unable to produce insulin
D	develops later in life	cells less sensitive to insulin

33 The diagram represents some neurons and the directions of impulses in a neural pathway.

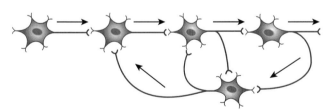

Which row in the table identifies the type of neural pathway shown and gives an example of a function in which it is involved?

	Type of neural pathway	Example of function
A	diverging	controlling breathing
B	diverging	controlling speech
C	reverberating	controlling speech
D	reverberating	controlling breathing

34 Which row in the table identifies a pair of antagonistic effects of the autonomic nervous system (ANS)?

	Sympathetic effect	Parasympathetic effect
A	decreased release of intestinal secretions	increased release of intestinal secretions
B	decreased heart rate	increased heart rate
C	increased peristalsis	decreased peristalsis
D	decreased breathing rate	increased breathing rate

35 The diagram shows a vertical section of the human brain.

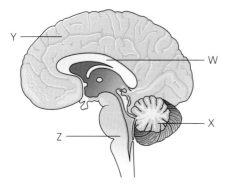

Which row in the table matches two neural functions with the areas labelled?

	Neural functions	
	Conscious thought	Control of heart rate
A	Y	X
B	W	Z
C	W	X
D	Y	Z

36 The corpus callosum
- **A** is the centre of conscious thought
- **B** transfers information between the cerebral hemispheres
- **C** receives nerve impulses from the sense organs
- **D** sends nerve impulses to the muscles and glands.

PRACTICE QUESTIONS

37 Students were asked to recall 12 words in any order, after hearing the list of words read out slowly. An analysis of their performance is shown in the graph.

The investigation was then repeated but with a delay before the students were allowed to write down the words that they could recall.

Which letter indicates the section of the graph that would be most likely to be affected by the delay?

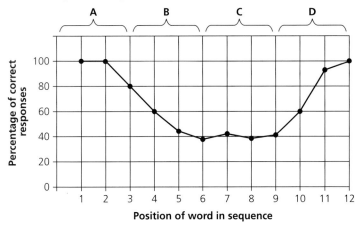

38 The diagram represents the flow of information through memory.

Which letter indicates the process that could be aided by the use of contextual cues?

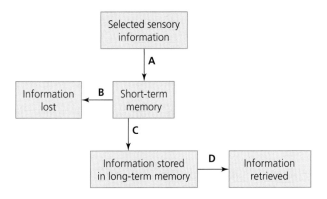

39 The diagram shows a motor neuron.

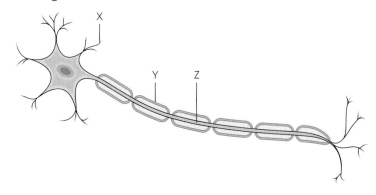

Which row in the table identifies the structures indicated by X, Y and Z?

	X	Y	Z
A	axon	myelin sheath	dendrite
B	dendrite	cell body	axon
C	dendrite	myelin sheath	axon
D	axon	cell body	dendrite

40 Some drugs used to treat neurotransmitter-related disorders are antagonists. These drugs

A block the action of the neurotransmitter

B mimic the action of the neurotransmitter

C inhibit the reuptake of the neurotransmitter

D inhibit the degradation of the neurotransmitter.

41 The diagram shows the action of a white blood cell.

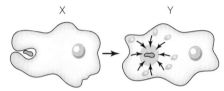

Which row in the table identifies the processes represented by X and Y in the diagram?

	X	Y
A	a phagocyte engulfing a pathogen	digestion of the pathogen with the help of the nucleus
B	a lymphocyte engulfing a pathogen	digestion of the pathogen with the help of antibodies
C	a phagocyte engulfing a pathogen	digestion of the pathogen with the help of lysosomes
D	a lymphocyte engulfing a pathogen	digestion of the pathogen with the help of lysosomes

42 Which of the following cells release histamine during the inflammatory response?

A phagocytes

B T lymphocytes

C mast cells

D B lymphocytes

43 Antibodies are proteins made by

A phagocytes that are specific to one antigen

B B lymphocytes that are specific to one antigen

C B lymphocytes that are non-specific

D phagocytes that are non-specific.

PRACTICE QUESTIONS

44 Failure of the regulation of the immune system leading to an autoimmune disease is caused by
- A   T lymphocytes responding to self-antigens
- B   B lymphocytes responding to self-antigens
- C   T lymphocytes responding to non-self-antigens
- D   B lymphocytes responding to non-self-antigens.

45 An adjuvant included in a vaccine is
- A   an inactivated toxin from a pathogen
- B   a weakened pathogen
- C   a chemical that inhibits the immune response
- D   a substance that enhances the immune response.

46 The influenza virus can evade the specific immune response by
- A   depleting lymphocytes
- B   invading phagocytes
- C   attacking memory cells
- D   showing antigenic variation.

47 In a clinical trial of a new vaccine, volunteers were split into two groups: A and B. Each group contained individuals matched on their age profiles.

Group A was given injections of the new vaccine and Group B was given injections of a dilute sugar solution.

Which of the following protocols is being described?
- A   placebo control
- B   family history analysis
- C   double-blind protocol
- D   randomised treatments

48 In a double-blind trial
- A   the participant, but not the investigator, knows who is getting the treatment under trial
- B   both the participant and the investigator know who is getting the treatment under trial
- C   the investigator, but not the participant, knows who is getting the treatment under trial
- D   neither the participant nor the investigator know who is getting the treatment under trial.

49 A DNA molecule contains 16 000 nucleotides, of which 30% contain thymine.

How many nucleotides in this molecule contain cytosine?
- A   1600
- B   2400
- C   3200
- D   6400

50 After a few seconds in a thermal cycling polymerase chain reaction (PCR) machine, a DNA sequence has been amplified to give 64 copies.

How many more cycles of PCR would be needed to amplify the sequence to 2048 copies?
- A   4
- B   5
- C   6
- D   11

51 Liver tissue contains an enzyme that breaks down alcohol. The graph shows the effect of copper ions on the rate of breakdown of alcohol by this enzyme over a 30-minute period.

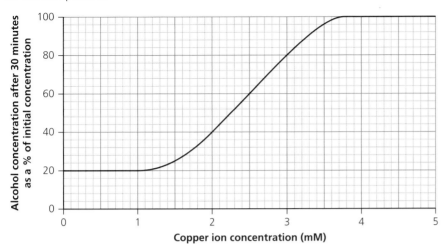

Which of the following conclusions can be drawn from the graph?

A 4.5 mM copper has no effect on enzyme activity.
B 2.5 mM copper halves the enzyme activity.
C 0.5 mM copper completely inhibits enzyme activity.
D Enzyme activity increases when copper ion concentration is increased from 1 mM to 3 mM.

52 In an experiment to show the effects of ATP on muscle fibre contraction, pieces of muscle measuring 50 mm were placed into 1% ATP solutions and their lengths measured after 5 minutes of immersion.

Which of the following would make the best control for this experiment?

Repeat the experiment but with pieces of muscle fibre

A in distilled water
B in glucose solution
C in 2% ATP solution
D left out of any solution.

53 The average durations of diastole and systole in a hospital patient over a period of time were measured and are shown below.

Diastole = 0.35 seconds

Atrial systole = 0.15 seconds

Ventricular systole = 0.30 seconds

What was the average heart rate of the individual over this period of time?

A 48 beats per minute
B 70 beats per minute
C 75 beats per minute
D 80 beats per minute

PRACTICE QUESTIONS

54 An experiment was carried out as shown in the diagram to compare the rates of respiration in a human athlete at different exercise intensities in a thermostatically controlled respirometer chamber. Different intensities of exercise were achieved by increasing the angle of the treadmill.

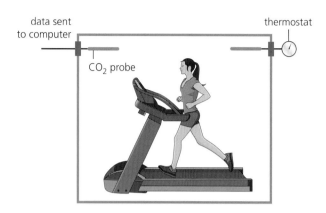

To improve the reliability of the results obtained from this experiment, it could be repeated

A over a range of temperatures

B at each exercise intensity

C with different starting air compositions

D using an oxygen probe.

Selecting

55 The table shows the percentage success rate of fertility treatment and the percentage of multiple births in women up to the age of 40.

14 000 women aged between 35 and 37 were given fertility treatment.

	Age of women receiving fertility treatment		
	Below 35	35–37	38–40
Percentage of fertility treatments resulting in live births	28	24	18
Multiple births as a percentage of the live births	26	20	18

How many multiple births were recorded from these women?

A 3360

B 2800

C 672

D 560

Processing

56 The graph shows the relationship between height and body mass and can be used to determine whether a person is underweight or overweight.

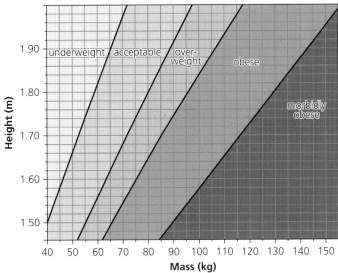

Which of the individuals shown in the table has a body mass index (BMI) that suggests they are overweight?

Individual	Mass (kg)	Height (m)
A	55	1.52
B	60	1.60
C	70	1.70
D	80	1.78

57 The effect of a virus infection on the antibody concentration in a patient's blood was investigated. The concentration of the antibody in their blood following the infection was measured and is shown in the table.

Time (days)	1	4	7	10	14	18	21	25
Antibody concentration (mg/100 cm³ of blood)	5.0	6.1	10.2	35.9	55.9	50.0	47.9	33.1

What was the average increase in antibody concentration per day (in mg/100 cm³) between day 10 and day 14?

A 5
B 10
C 20
D 50.4

PRACTICE QUESTIONS

58 In an investigation, the mass of myelin in brain tissue was measured in a control group without dementia and in patients with different types of dementia as shown.

Group 1 without dementia
Group 2 vascular dementia
Group 3 Alzheimer dementia
Group 4 Lewy dementia

The results are shown in the bar chart.

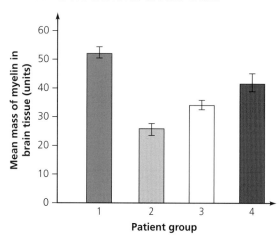

Which conclusion can be drawn from these results?

A   Loss of myelin from the brain causes dementia.
B   Lewy is the most severe form of dementia.
C   Most myelin loss is linked with vascular dementia.
D   Alzheimer dementia is more severe than Lewy dementia.

Concluding

59 The table shows the number of cases of influenza in a health board area over a 5-year period.

Year	Influenza cases in January	Influenza cases in July
2014	580	120
2015	620	345
2016	1200	350
2017	120	145
2018	400	100

Which conclusion can be drawn from the information in the table?

A   There are always more cases of influenza in January than in July.
B   The number of influenza cases in January increased steadily from 2014 until 2018.
C   More people died of influenza in 2016 than in any other year.
D   The number of cases of influenza decreased by 75% between January and July 2018.

Concluding

60 During a set of trials to investigate the serial position effect, a group of participants was read a list of 12 letters. Ten seconds later they were asked to recall the list in its original order. Their responses were analysed and the results shown in the bar chart.

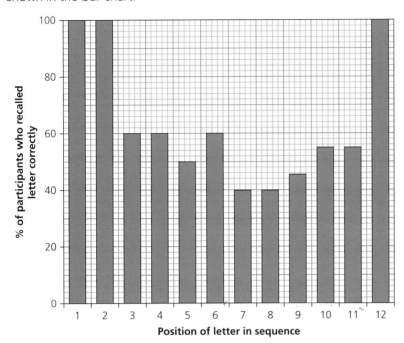

What percentage of the letters were correctly recalled in their correct positions by 50% or more of the participants?

A 9
B 25
C 50
D 75

PRACTICE QUESTIONS

Question type: Structured and extended response

Paper 2 of your examination is made up of structured and extended response questions for a total of 95 marks. Again, a mark should take about 1.5 minutes but questions with lots of reading or thinking time or those with calculations or information to process will take longer and some straightforward questions can be done more quickly.

In all cases you should pay careful attention to the mark allocation and the number of answer lines or space provided. Each individual mark is awarded separately, so if a question is worth 2 marks there will be two parts to the answer required. If several answer lines are provided, you will probably need to use them.

For the practice questions given here, you should write your answers on a separate piece of paper.

The structured questions are of different types: short answer, data handling and experimental, but these will be mixed together in the actual examination paper and there is some overlap.

There are two types of extended response questions: mini and full.

> **Top Tip!**
> Remember, Paper 2 of your examination should take no more than 2 hours and 20 minutes.

≫ HOW TO ANSWER

Short-answer questions

Most of the short-answer questions in your examination are focused on testing knowledge. They are often introduced by a short sentence about a Key Area and it is very common to have a labelled diagram presented here. There are likely to be 3–5 marks available for related answers. Very occasionally you may be given a choice of question.

Many of the questions will be at the demand level for Grade C, where you need to name, state or give answers or identify structures. These questions test your memory of the Key Areas (in other words, Demonstrating KU – demonstrating knowledge and understanding).

> **Top Tip!**
> Questions that ask for descriptions, explanations or suggestions are often worth multiple marks, so remember to give a statement for each mark.
>
> 'Explain' questions will always require you to bring in correctly selected additional knowledge. There may be several acceptable answers to 'Suggest' questions.

Some of the questions will be at the demand level for Grade A, which often ask for descriptions, explanations or suggestions. These questions are testing your understanding of Key Areas (in other words, Applying KU – application of knowledge and understanding). There are also likely to be some Skills of Scientific Inquiry (SSI) marks mixed in with these questions.

Make sure that you revise your Key Areas thoroughly and systematically. Be aware that the words and terms sought in the answers are those that are given in the Course Specification for Human Biology – this is crucial.

PRACTICE QUESTIONS

1 a The table shows some information about cell types and cell division in the body.
Copy and complete the table by adding the appropriate terms to the boxes.

Parent cell type	Type of cell division	Daughter cell type
somatic		
germline		gamete

b Give **one** use of stem cells in medical research.

2 a During replication, the two strands of a DNA molecule separate and each acts as a template for the production of a new strand.

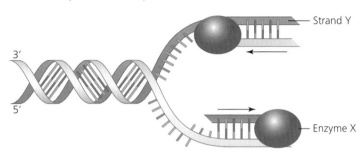

  i   Name enzyme X.
  ii Identify the feature, shown in the diagram, which confirms that strand Y is the leading strand.
  iii State the structural difference between the 3′ and 5′ end of a DNA strand.

b Explain why DNA replication must take place before a cell divides.

3 The diagram shows the synthesis of a polypeptide in a cell.

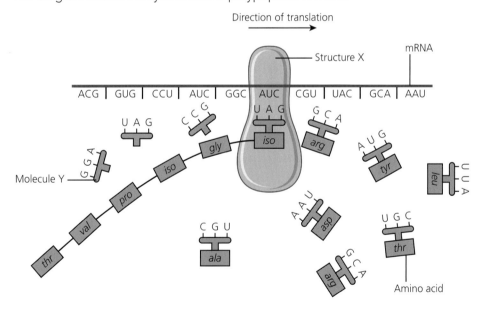

a Name structure X.
b In the diagram, the amino acid *iso* has just been added to the polypeptide chain.
Complete the boxes to identify the missing amino acids that will be added as the polypeptide chain is completed.

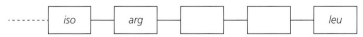

c Give the DNA triplet that codes for the amino acid shown as *asp*.

PRACTICE QUESTIONS

4 The diagram shows a gene mutation and the effects of this mutation on the mRNA produced when the gene is transcribed.

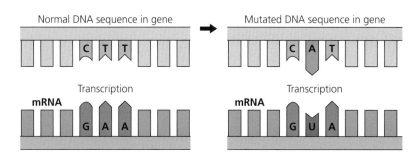

- **a** i Name the type of gene mutation that has occurred in this example. [1 mark – Applying KU]
  - ii Describe the effect this mutation would be expected to have on the amino acid chain produced when the mRNA is translated. [1 mark – Applying KU]
- **b** Cri du chat syndrome results from missing a piece of chromosome 5.
  Name this type of chromosome mutation. [1 mark – Applying KU]
- **c** Describe what is meant by a translocation mutation. [1 mark – Demonstrating KU]

5 a Describe what is meant by the 'genome' of an organism. [2 marks – Demonstrating KU]
  b Explain how the analysis of individual genomes may lead to personalised medicine. [2 marks – Demonstrating KU]

6 The diagram shows four metabolites A–D in a metabolic pathway.

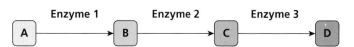

- **a** i Describe how the concentration of metabolite D could be controlled by feedback inhibition. [1 mark – Applying KU]
  - ii Describe the effect on the concentration of metabolites B and C if a non-competitive inhibitor of enzyme 2 is added. [2 marks – Applying KU]
- **b** Describe the role of genes in the control of metabolic pathways such as this. [1 mark – Demonstrating KU]

PRACTICE QUESTIONS

7 The diagram shows parts of metabolic pathways in cellular respiration within a skeletal muscle cell.

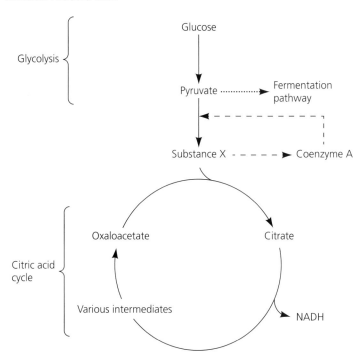

- **a** Name substance X. — 1 — Applying KU
- **b** Describe the role of the coenzyme NAD. — 1 — Demonstrating KU
- **c** Under certain conditions, pyruvate enters a fermentation pathway, as shown. Describe the conditions under which this might occur and name the substance into which the pyruvate would be converted. — 2 — Applying KU

8 a The diagram represents glycolysis and the metabolic pathway that occurs during vigorous exercise.

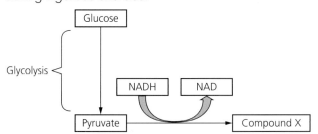

  i Name compound X. — 1 — Applying KU
  ii Explain the importance to glycolysis of the regeneration of NAD shown in the diagram. — 1 — Applying KU

b Athletes show distinct patterns of muscle fibres that reflect their sporting activities.

Name a sporting activity for which a higher percentage of slow-twitch muscle fibres could confer an advantage and describe one feature of these fibres. — 2 — Applying KU

PRACTICE QUESTIONS

9 The flowchart summarises some of the processes involved in the male reproductive system.

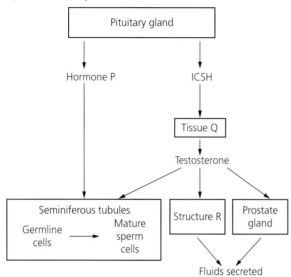

- **a** Name the type of cell division that results in the production of gametes from germline cells. **1**
- **b** Name hormone P and tissue Q. **2**
- **c** Name structure R, which produces fluids to maintain the mobility and viability of the sperm. **1**

10 a The flowchart shows the system controlling sperm production and the concentration of testosterone.

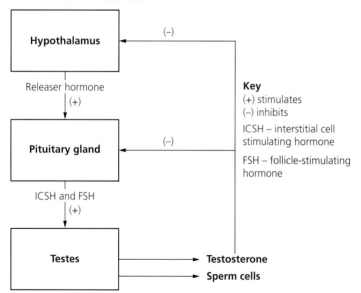

Explain how the concentration of testosterone in the blood is prevented from becoming too high. **2**

b The graph shows the concentration of two ovarian hormones in a woman's blood during one menstrual cycle.

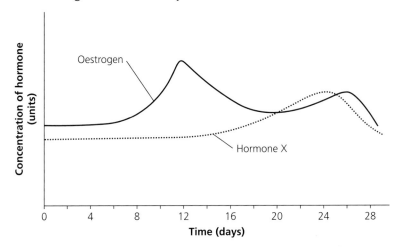

   i Name hormone X. — 1 — Applying KU
   ii Describe **one** way in which the graph would be different if the woman became pregnant during this cycle. — 1 — Applying KU

11 a Female fertility is cyclical.
Describe how the fertile period of a woman can be calculated. — 2 — Demonstrating KU

b Nexplanon is the female contraceptive implant used in the UK.
The implant is a small flexible tube inserted under the skin of the upper arm under local anaesthetic. It releases a synthetic form of progesterone.
Explain how Nexplanon prevents fertilisation from taking place. — 2 — Applying KU

12 a Where conditions such as cystic fibrosis exist in a family, the family history can be used to determine the genotypes of its individual members.
   i State the term used for this process. — 1 — Demonstrating KU
   ii Give **one** characteristic of a family history chart that would enable a geneticist to establish that the condition or disorder was autosomal recessive. — 1 — Demonstrating KU
   iii Give **one** characteristic of a family history chart that would enable a geneticist to establish that the condition or disorder was sex-linked and recessive. — 1 — Demonstrating KU

b During IVF treatment, it is possible to detect single gene disorders in fertilised eggs before they are implanted into the mother.
Give the term that describes this procedure. — 1 — Demonstrating KU

13 The diagram shows a cross-section through a vein in an individual's leg.

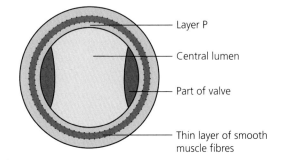

a Name layer P, which lines the central lumen of the vein. — 1 — Applying KU
b Describe the role of valves in veins. — 1 — Applying KU
c **i** State what is meant by deep vein thrombosis (DVT). — 1 — Demonstrating KU
   ii Describe **one** effect that DVT can have on the body. — 1 — Demonstrating KU

PRACTICE QUESTIONS

14 The diagram shows the cardiac conducting system in the human heart. The arrows show the pathways taken by the electrical activity involved in co-ordinating the heartbeat in the cardiac cycle.

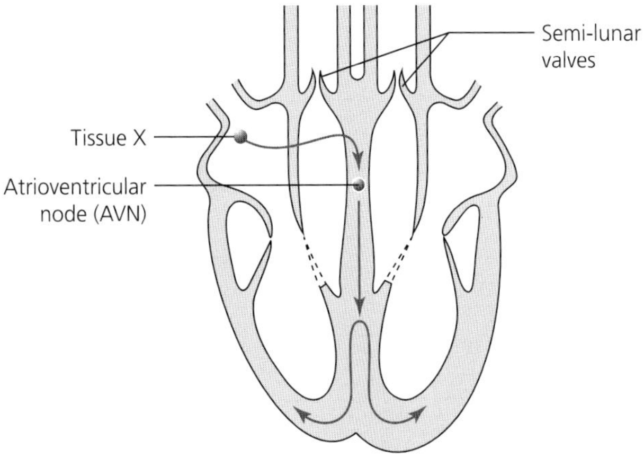

	MARKS	STUDENT MARGIN
a Name the region of nervous tissue labelled X.	1	Applying KU
b When a wave of electrical activity reaches the AVN, there is a short delay before a new wave leaves the AVN. Explain the importance of this short delay.	1	Applying KU
c Describe the state of the heart valves during ventricular systole.	2	Demonstrating KU

15 Atherosclerosis is a condition that can cause hypertension. It can lead to similar effects to a clot in the coronary arteries and also increases the risk of stroke.

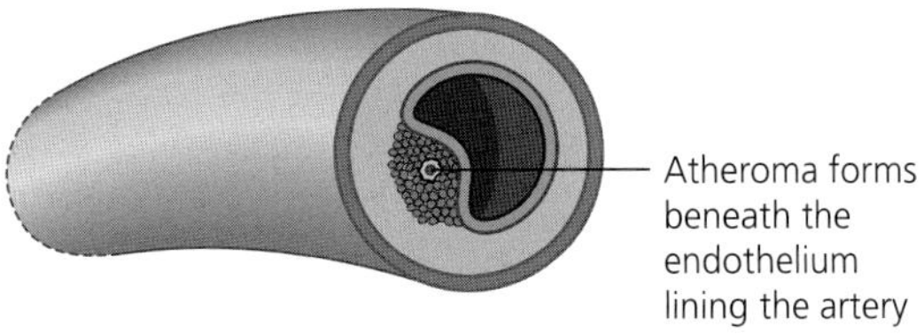

Atheroma forms beneath the endothelium lining the artery

	MARKS	STUDENT MARGIN
a Give **two** effects that atherosclerosis has on the structure of an artery.	2	Demonstrating KU
b Atheromas may rupture, damaging the endothelium. This releases clotting factors that activate a cascade of reactions leading to the formation of a clot.		
i Name the activated form of the enzyme involved in blood clotting.	1	Demonstrating KU
ii Name the soluble form of the blood protein that helps to form a blood clot.	1	Demonstrating KU

16 a The diagram shows some stages in the control of blood glucose concentration.

Stage 1 Decrease in blood glucose concentration below normal is detected by receptors in organ P

↓

Stage 2 Increased secretion of hormone Q

↓

Stage 3 Liver converts storage carbohydrate to glucose

↓

Stage 4 Blood glucose concentration returns to normal

↓

Stage 5 Corrective mechanism switched off

	MARKS	STUDENT MARGIN
i Name organ P.	1	Applying KU
ii Identify hormone Q.	1	Applying KU
iii Name the carbohydrate stored in the liver.	1	Demonstrating KU

PRACTICE QUESTIONS

b A pancreas transplant is an operation to treat type 1 diabetes by replacing the need for insulin with a healthy insulin-producing pancreas from a donor.

Explain why pancreas transplants are not used for the treatment of type 2 diabetes. **(1)** *Applying KU*

17 Sympathetic and parasympathetic nerves regulate peristalsis.

   a **i** Name the part of the brain that regulates peristalsis. **(1)** *Demonstrating KU*

      ii Give the term that describes the relationship between sympathetic and parasympathetic nerves. **(1)** *Demonstrating KU*

   b The diagram shows a neural pathway in the nervous system.

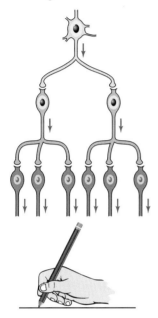

      i Name this type of neural pathway. **(1)** *Applying KU*

      ii Describe the action of this neural pathway. **(1)** *Demonstrating KU*

18 The diagram shows a section through the brain.

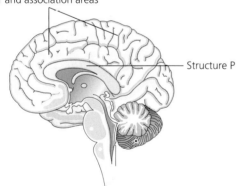

Cerebral cortex containing sensory, motor and association areas

Structure P

   a Give the function of motor areas of the cerebral cortex. **(1)** *Demonstrating KU*

   b Give **one** example of a function of an association area of the cerebral cortex. **(1)** *Applying KU*

   c Identify structure P and describe its function. **(2)** *Applying KU*

PRACTICE QUESTIONS

19 The flowchart shows some processes involved in memory.

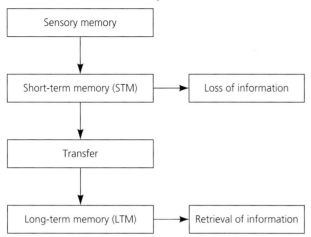

a Explain why loss of information from the short-term memory can occur. **1** *Demonstrating KU*

b One method of transferring information to the long-term memory, as shown in the flowchart, involves elaborative encoding.

   i Give **one** other method of information transfer. **1** *Demonstrating KU*

   ii Describe how elaborative encoding differs from shallow encoding. **1** *Demonstrating KU*

c Give the name of the model that is used to explain why the short-term memory can perform simple cognitive tasks. **1** *Demonstrating KU*

20 The diagram shows a neuron from an adult.

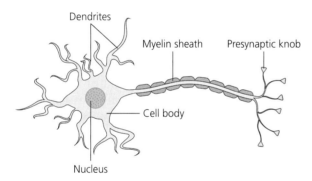

a   i Suggest a role of the nucleus in the transfer of information across a synapse. **1** *Applying KU*

  ii Describe how a neuron in a new-born child could differ from the neuron shown in the diagram and how this would affect how the neuron functions. **1** *Applying KU*

  iii Name the type of cells responsible for the production of the myelin sheath. **1** *Demonstrating KU*

b Neurotransmitters are removed from the synaptic cleft immediately following the passing on of any impulse.

Describe **one** method by which removal of the neurotransmitter is achieved. **1** *Applying KU*

PRACTICE QUESTIONS

21 The diagram shows cells in a region of the skin that has been damaged through accidental piercing by a metal pin.

The flowchart shows some of the events that can result from this type of damage to skin.

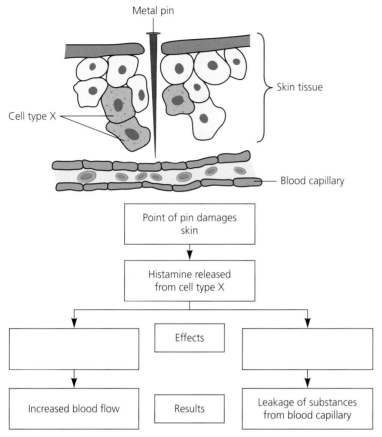

a Identify cell type X. — 1 — Applying KU

b Copy and complete the flow chart to show the effects of histamine release. — 2 — Applying KU

c The phagocytes release small proteins called cytokines.
Describe the role of the cytokines in cellular defence. — 1 — Demonstrating KU

22 The diagram shows how a cell in an individual's immune system responds to a virus.

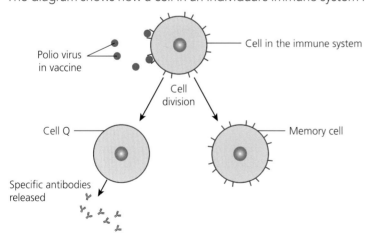

a i Name the substances present on the surface of the virus that trigger an immune response. — 1 — Applying KU

ii Name cell Q. — 1 — Applying KU

iii Describe the specificity of the antibodies shown in the diagram. — 1 — Applying KU

b Name the type of white blood cell that induces apoptosis in cells that have become infected with the virus. — 1 — Demonstrating KU

29

PRACTICE QUESTIONS

			MARKS	STUDENT MARGIN
23	The best way to protect an individual against a disease is by direct vaccination rather than by relying on 'indirect' protection through herd immunity.			
	a	i Describe what is meant by herd immunity.	1	Demonstrating KU
		ii Give **one** factor that influences the threshold percentage of immune people needed to achieve effective herd immunity.	1	Demonstrating KU
	b	i Give **one** source of antigen added to a vaccine.	1	Demonstrating KU
		ii State why vaccines usually contain an adjuvant.	1	Demonstrating KU
24	New vaccines are subject to clinical trials before being licensed for use.			
	a	State what is meant by a clinical trial.	1	Demonstrating KU
	b	Describe **two** design features of clinical trials.	2	Demonstrating KU
	c	Explain the importance of group size in clinical trials.	1	Demonstrating KU

≫ HOW TO ANSWER

Data-handling questions

These questions focus on the results of an experiment or investigation that has been carried out. There will be one large data-handling question in Paper 2 of your examination, but some of the individual skills can be tested within the short-answer questions too.

Results data will be given as either a table or a graph. You can be asked to draw a line graph or bar chart if the results are presented as a table.

You may have to read the data and select specific values. There will very often be calculations to do – average, average change, ratio or percentage change are the most common, but you can be asked to do +, −, × and ÷ sums as well.

Top Tip!
Draw graphs and bar charts using a ruler and use the data table headings and units for the axis labels. Marks are given for providing scales, labelling the axes correctly and plotting the data points or drawing the bars. Line graphs require points to be joined with straight lines using a ruler.

Top Tip!
If you are given space for calculation, you will very likely need to use it!

Top Tip!
Remember to use values from the graph when describing graph trends if you are asked to do so.

If you are asked to calculate an increase or decrease between points on a graph, you should use a ruler to help accuracy – draw pencil lines on the actual graph if this helps.

You might be asked to predict results for situations not tested – this could be higher or lower temperature, pH or light intensity, for example. Predictions can involve extending graph trends or reading between values on a table.

Make sure you have a sharp pencil, a ruler and a calculator to help with these questions. A highlighter pen can also help.

PRACTICE QUESTIONS

1. As part of an investigation into metabolic rate and work rate, an elite athlete with a body mass of 80 kg performed various intensities of exercise on a treadmill. His oxygen consumption and pulse rate were measured and the results shown in the table. The graph shows the relationship between oxygen consumption and work rate.

   Table

Intensity of exercise (units)	Description	Oxygen consumption (cm^3 per minute per kg)	Pulse rate (bpm)
1	standing	3.5	40
2	walking	10	42
3	jogging	20	44
4	running	30	70
5	sprinting	70	100

   Graph

   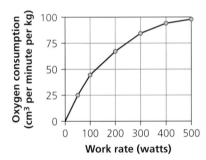

   a. Apart from the measurement of oxygen consumption, state **one** other measurement which could be used to estimate an individual's metabolic rate. **[1 — Planning]**

   b. From the table
      i. calculate the ratio of oxygen consumed when standing compared to sprinting **[1 — Processing]**
      ii. calculate the volume of oxygen that would be consumed by the athlete when jogging for 5 minutes. **[1 — Processing]**

   c. Describe how increasing exercise intensity affected the pulse rate of the athlete. **[2 — Selecting]**

   d. Using the data provided, predict
      i. the category of activity being performed by the athlete when their work rate was 200 watts **[1 — Predicting]**
      ii. the athlete's work rate when they are consuming 50 cm^3 of oxygen per minute. **[1 — Selecting]**

   e. Calculate the percentage increase in work rate when oxygen consumption rises from 25 cm^3 to 75 cm^3 per minute per kg. **[1 — Processing]**

2. a. Body mass index (BMI) is a measurement of body fat based on height and mass, as shown in the equation.

   $$BMI = \frac{body\ mass\ (kg)}{height\ (m)^2}$$

   It can be used to identify those who are overweight, obese, underweight or healthy.

   The table on the following page shows the BMI measurements used to categorise individuals.

   The graph shows the changes in body mass and height of a male from the age of 1 to 21 years.

PRACTICE QUESTIONS

BMI range	Category
less than 18.5	underweight
18.5–24.9	healthy
25–29.9	overweight
30+	obese

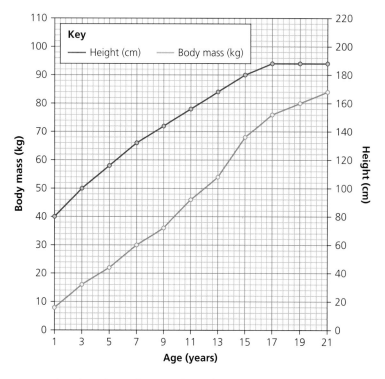

i Use the data from the graph to describe the changes that occur in the height of this male between the age of 1 and 21. **2** Selecting

ii Calculate the average yearly increase in body mass between 7 and 17 years of age. **1** Processing

iii Calculate the BMI of this male at the age of 21 and give the category that this relates to. **1** Processing

iv Another individual of height 1.8 m has a BMI of 20. Calculate this individual's body mass. **1** Processing

b The bar chart shows the link between BMI and the relative risk of developing diabetes. A relative risk of 1 is the normal risk of developing diabetes.

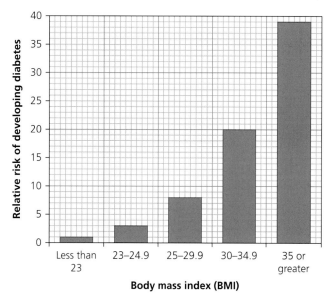

i Give a conclusion that can be made from the information in the bar chart.

ii Express, as a simple whole number ratio, the relative risk of developing diabetes of individuals with a BMI of 35 or greater compared to those with a BMI between 23 and 24.9.

iii Using information in the table and the bar chart, calculate the percentage increase in the relative risk of developing diabetes if an individual's BMI changed from being described as underweight to being overweight.

3 Receptors on the membranes of neurons are activated by natural neurotransmitters and by agonistic drugs, which mimic neurotransmitter action. Activation of these receptors produces an electrical response by the neuron.

Graph 1 shows the results of an investigation into the effects of the concentration of the agonistic drugs morphine and buprenorphine on the electrical response of neurons.

Graph 1

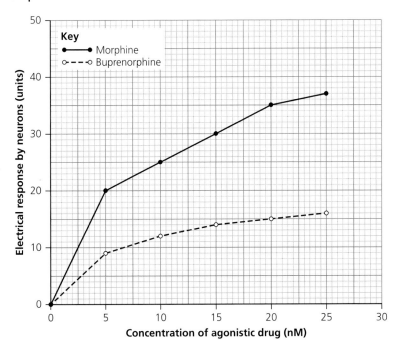

Antagonistic drugs can block the effects of neurotransmitters and agonistic drugs.

PRACTICE QUESTIONS

Graph 2 shows how neurons treated with a 3 nM solution of morphine responded to increasing concentration of an experimental antagonistic drug.

Graph 2

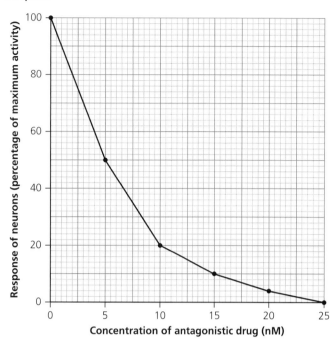

a   Using information in **Graph 1** on the previous page
    i   calculate the difference in response between the use of 15 nM morphine and 15 nM buprenorphine **1** Processing
    ii calculate the percentage increase in response when the concentration of morphine was increased from 5 nM to 10 nM **1** Processing
    iii describe the effect of increasing the concentration of morphine on the electrical response by neurons **2** Selecting
    iv predict the electrical response of neurons if they were exposed to 30 nM of morphine. **1** Predicting
b   Using information in **Graph 1** and **Graph 2**, calculate the electrical activity in the neurons which had been exposed to 3 nM morphine and 5 nM of the experimental antagonistic drug. **1** Processing
c   Using information in **Graph 2**, give the concentration of the antagonistic drug needed to reduce the neuron response by 70%. **1** Selecting
d   Calculate the average decrease in neuron response for every 5 nM of the antagonistic drug. **1** Processing

≫ HOW TO ANSWER

Experimental questions

These questions focus on an experiment or investigation that has been carried out. There will be one large experimental question in Paper 2 of your examination, but some of the individual skills can be tested within the short-answer questions too.

Often there is a very short description of the experimental method, sometimes with a diagram of the apparatus used, followed by some results data given as a table or a graph.

> **Top Tip!**
> Try to bring all the information together in your mind to visualise what has been done – doing a small pencil drawing in the margin can sometimes help.

> **Top Tip!**
> Any variable that could affect the results of an experiment should be controlled. A control experiment allows a comparison to be made and allows you to relate the dependent variable to the independent one. The control should be identical to the original experiment apart from the one factor being investigated.

These questions can ask for evaluation of, or comments about, planning – such as controls, validity or reliability.

They can also lead into questions about the experimental data, like those described in data-handling questions, such as drawing graphs.

You should be able to identify the independent and dependent variables in an investigation or experiment.
- ▶ The independent variable is the input variable – it usually appears in the first column of a data table and is plotted on the x-axis of a graph.
- ▶ The dependent variable refers to the data (results) produced – it usually appears in the second column of a data table and is plotted on the y-axis of a graph.

You may also be asked to draw valid conclusions, giving explanations supported by evidence.

When concluding, you must refer to the experimental aim, which is likely to be stated in the stem of the question.

Make sure you have a sharp pencil, a ruler and a calculator to help with these questions. A highlighter pen can also help.

PRACTICE QUESTIONS

1 Patients requiring an organ transplant are tissue typed to match with potential donors. Polymerase chain reaction (PCR) is used to amplify DNA from a patient and from potential donors. Gel electrophoresis is used to compare DNA sequences of the patient with those of donors. The presence of specific DNA bands in the gel can indicate that a donor is a suitable match.

In a procedure, amplified DNA from a patient and three potential donors was compared by gel electrophoresis as shown in the diagram. The table shows how the size of a DNA fragment affects the distance it travels in the gel.

Diagram

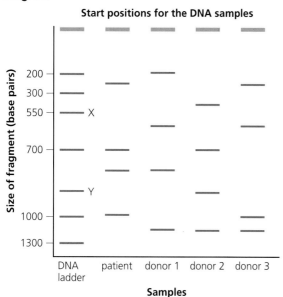

Table

Size of DNA fragment (bp)	Distance travelled (mm)
200	72
300	58
550	32
700	18
800	12
1000	10
1300	8

a Identify **one** variable that must be controlled when carrying out gel electrophoresis.

b Suggest what should be done to increase the reliability of the procedure.

c Identify the distance travelled by fragment X.

d On a separate piece of graph paper, construct a line graph to show how the size of a DNA fragment affects how far it travels in the gel.

e Describe the relationship between the size of a DNA fragment and the distance it travels in the gel.

f Fragment Y has travelled 11 mm in the gel.

Give the number of base pairs in this fragment.

g Identify the donor whose DNA best matches that of the patient.

PRACTICE QUESTIONS

| | | MARKS | STUDENT MARGIN |

2 As part of a series of clinical trials, the systolic and diastolic blood pressures of six young adult participants were measured. Each participant was asked to drink 500 cm³ of an energy drink and their blood pressure was measured again one hour after taking the drink.

The results are shown in the table.

Participant	Initial blood pressure (mm Hg)		Blood pressure one hour after taking the drink (mm Hg)	
	Systolic	Diastolic	Systolic	Diastolic
1	120	75	134	82
2	127	80	145	84
3	118	70	124	72
4	134	81	143	83
5	122	73	133	77
6	129	83	137	88
Average reading	125	77	136	81

a Calculate the percentage increase in the diastolic pressure of participant 2 one hour after taking the energy drink. — 1 — Processing

b **i** Identify the dependent variable in this investigation. — 1 — Planning

 ii Identify **one** variable, not already stated, which would have to be kept constant during this investigation. — 1 — Evaluating

c Describe an appropriate control for this investigation. — 1 — Planning

d On a separate piece of graph paper, construct a bar chart to show all the average blood pressure readings shown in the table. — 2 — Presenting

e **i** Give **one** conclusion that could be drawn from the results in the table. — 1 — Concluding

 ii Suggest why conclusions from these results might not be valid. — 1 — Planning

3 During a set of trials to investigate the serial position effect, two groups of ten participants were read a list of 15 words without pauses. Ten seconds later they were asked to recall the list. The results are shown in the table.

Position of the word in the series	Percentage of correct responses
1	100
2	90
3	100
4	90
5	80
6	60
7	50
8	40
9	50
10	60
11	70
12	80
13	100
14	90
15	100

37

PRACTICE QUESTIONS

			MARKS	STUDENT MARGIN
a	i	On a separate piece of graph paper, construct a line graph to show the data in the table.	2	Presenting
	ii	Describe the relationship between the position of the word in the series and the percentage of correct responses.	2	Selecting
	iii	Identify **one** feature of this investigation that increases the reliability of the results.	1	Planning
	iv	Calculate the number of students who correctly recalled the words at position 8.	1	Processing

b A second investigation was carried out which included a delay of 30 seconds before students were asked to recall the list of words. The results are shown in the table.

Position of the word in the series	Percentage of correct responses
1	100
2	90
3	100
4	90
5	80
6	60
7	50
8	40
9	50
10	60
11	50
12	60
13	60
14	50
15	60

			MARKS	STUDENT MARGIN
	i	Describe the effect of the delay on the ability of the participants to recall the words in the list.	1	Selecting
	ii	Give **one** variable that would need to be controlled to allow a valid conclusion to be made.	1	Planning

PRACTICE QUESTIONS

❯❯ HOW TO ANSWER

Mini extended response questions

There will usually be a maximum of two open-ended questions in Paper 2 of your examination paper, for which you will need to give extended responses. There could be a choice of question, but not always.

Each question will be short but several answer lines will be given, which will be a good clue to the answer length. You will need several sentences for a full answer in each case, and there will be 3–5 marks available for each question. Each mark is awarded separately, so the mark allocation gives a clue to the expected answer length too.

> **Top Tip!**
> Read the question very carefully. If there is a choice, be clear about which you are selecting.

> **Top Tip!**
> 'Give an account of' means the same as 'describe'.

The questions test the understanding of related knowledge, so you could be asked to describe a process or to compare structures or processes. If you are asked to describe a process, remember to be logical, starting from the beginning of the process and working through in steps. If you are asked to compare two processes or structures, ensure that you describe both of them in full.

		MARKS	STUDENT MARGIN
1	Compare embryonic and tissue stem cells.	5	Demonstrating KU
2	Describe the main steps in the polymerase chain reaction (PCR).	4	Demonstrating KU
3	Give an account of lactate metabolism in muscle cells.	5	Demonstrating KU
4	Explain the biological basis for the stimulation of ovulation by ovulatory drugs.	5	Demonstrating KU
5	Describe the antagonistic action of the autonomic nervous system in the control of the cardiac cycle.	4	Demonstrating KU
6	Describe the process of atherosclerosis and its effect on arteries and blood pressure.	5	Demonstrating KU
7	Give an account of the structure and function of neural pathways.	4	Demonstrating KU
8	Give an account of long-term memory.	5	Demonstrating KU
9	Give an account of vaccination.	5	Demonstrating KU

PRACTICE QUESTIONS

⟫ HOW TO ANSWER
Full extended response questions

There will be one full extended response question at the end of Paper 2, which always contains a choice.

Top Tip!
Read the questions very carefully and spend a minute or two to decide which choice is best for you, in other words which one you can recall the most about.

Top Tip!
Each mark is awarded for a separate statement, so the mark allocation gives a clue to the expected answer length.

Some extended response questions are divided into two or even three parts. It's best to answer each part separately under its heading. There are 7–10 marks for this question and you will need to make a statement for each mark.

Like the mini extended response questions, these questions test the understanding of related knowledge, so you could be asked to describe a process or to compare structures or processes. If you are asked to describe a process, remember to be logical, starting from the beginning of the process and working through in steps. You are allowed to include diagrams in your answer, but ensure that these are labelled.

		MARKS	STUDENT MARGIN
1	Describe the structure and replication of a molecule of DNA.	9	Demonstrating KU
2	Describe the mode of action of enzymes in the control of metabolic pathways.	9	Demonstrating KU
3	Describe hormonal control of the menstrual cycle under the following headings:		
a	events leading to ovulation	5	Demonstrating KU
b	events following ovulation.	4	Demonstrating KU
4	Describe the cardiac cycle under the following headings:		
a	the conducting system of the heart	4	Demonstrating KU
b	nervous control of the cardiac cycle.	5	Demonstrating KU
5	Give an account of the nervous system under the following headings:		
a	the role of neurotransmitters at the synapse	5	Demonstrating KU
b	the structure and function of neural pathways.	4	Demonstrating KU
6	Give an account of immunisation under the following headings:		
a	vaccination	5	Demonstrating KU
b	herd immunity and the difficulties encountered in achieving widespread vaccination.	4	Demonstrating KU

ANSWERS TO PRACTICE QUESTIONS

Multiple-choice

Question	Answer	Mark	Demand
1	B	1	C
2	A	1	C
3	B	1	A
4	A	1	C
5	A	1	C
6	B	1	C
7	A	1	C
8	A	1	A
9	D	1	C
10	C	1	C
11	B	1	A
12	D	1	A
13	C	1	A
14	B	1	C
15	D	1	C
16	B	1	A
17	B	1	C
18	C	1	C
19	D	1	A
20	B	1	A
21	B	1	C
22	A	1	C
23	C	1	A
24	D	1	C
25	B	1	C
26	D	1	C
27	B	1	C
28	C	1	A
29	A	1	C
30	A	1	C
31	D	1	C
32	C	1	C
33	D	1	C
34	A	1	C
35	D	1	C
36	B	1	C
37	D	1	A
38	C	1	C
39	C	1	C
40	A	1	C
41	C	1	C
42	C	1	C

ANSWERS TO PRACTICE QUESTIONS

Question	Answer	Mark	Demand
43	B	1	C
44	A	1	C
45	D	1	C
46	D	1	C
47	A	1	C
48	D	1	C
49	C	1	A
50	B	1	C
51	B	1	A
52	A	1	C
53	C	1	A
54	B	1	C
55	C	1	A
56	D	1	C
57	A	1	A
58	C	1	A
59	D	1	C
60	D	1	C

Short-answer questions

Question			Expected answer	Mark	Demand
1	a		Somatic cells: mitosis = 1; somatic = 1 Germline cells: meiosis = 1	3	CCC
	b		Model cells to study how diseases develop **OR** for drug testing **OR** to study how cellular processes such as cell growth/differentiation/gene regulation work	1	C
2	a	i	DNA polymerase	1	C
		ii	The leading strand is replicated continuously **OR** DNA polymerase is adding nucleotides towards the unwinding DNA/replication fork	1	A
		iii	Deoxyribose is located at the 3´ **AND** phosphate is located at the 5´ end	1	A
	b		To ensure that each daughter cell receives all the genetic information needed to carry out all of its functions	1	C
3	a		Ribosome	1	C
	b		tyr – ala (1 mark each)	2	CC
	c		AAT	1	C
4	a	i	Substitution	1	C
		ii	(Change in one codon) causes a change in one amino acid	1	C
	b		Deletion	1	C
	c		Detached genes become attached to a non-homologous chromosome	1	A

ANSWERS TO PRACTICE QUESTIONS

Question			Expected answer	Mark	Demand
5	a		The entire hereditary information encoded in DNA = 1 The DNA sequences in genes and non-coding regions = 1	2	CA
	b		An individual's genome may be analysed = 1 Drug therapy/dosage can be matched to the genome to gain the most effective outcome from treatment = 1	2	CA
6	a	i	Build-up of D acts as an inhibitor of an earlier step in the pathway so leads to reduced concentration of D	1	A
		ii	B would increase = 1 C would decrease = 1	2	CC
	b		Genes encode the enzymes which control the pathway	1	A
7	a		Acetyl group	1	C
	b		Transfers/carries hydrogen ions/H+ **AND** electrons to the electron transport chain/inner membrane of the mitochondria	1	A
	c		Anaerobic/absence/shortage of oxygen/high levels of exercise = 1 Lactate = 1	2	CC
8	a	i	Lactate	1	C
		ii	Needed to maintain ATP production through glycolysis	1	A
	b		Named activity, e.g. long-distance running/cycling/cross-country skiing = 1 They have many blood vessels/more mitochondria/more myoglobin **Any 1** = 1	2	CA
9	a		Meiosis	1	C
	b		P = FSH = 1 Q = Interstitial cells = 1	2	CC
	c		Seminal vesicles	1	C
10	a		Overproduction of testosterone is prevented by a negative feedback control/mechanism = 1 High testosterone levels inhibit the secretion of FSH/ICSH from the pituitary gland (resulting in a decrease in the production of testosterone by the interstitial cells) = 1	2	CA
	b	i	Progesterone	1	C
		ii	Progesterone levels would remain high	1	C
11	a		Time of ovulation can be estimated based on a slight rise in body temperature/0.5 °C on the day of ovulation = 1 Thinning of cervical mucus = 1	2	CC
	b		Negative feedback = 1 High progesterone prevents the release of FSH/LH from the pituitary gland = 1	2	CA
12	a	i	Genetic screening	1	C
		ii	An autosomal recessive disorder is expressed relatively rarely in the offspring **OR** affects males and females equally **OR** may skip generations	1	C
		iii	In sex-linked recessive disorders, males are affected more often than females **OR** male offspring receive the condition from their mother	1	C
	b		Pre-implantation genetic diagnosis/PGD	1	C

ANSWERS TO PRACTICE QUESTIONS

Question			Expected answer	Mark	Demand
13	a		Endothelium	1	C
	b		Prevent the backflow of blood	1	C
	c	i	A blood clot which occurs within a vein (deep in the body tissues)	1	C
		ii	Slows or stops the passage of blood through the vein **OR** If a thrombus breaks loose, it forms an embolus that can travel through the bloodstream and block a blood vessel/pulmonary embolism	1	C
14	a		Sinoatrial node/SAN	1	C
	b		Time to allow the ventricles to fill with the blood from the atria/ventricular diastole	1	A
	c		Atrioventricular/AV valves are shut = 1 Semi-lunar valves are open = 1	2	CA
15	a		The artery thickens/loses its elasticity/reduces the diameter of the lumen **Any 2**	2	CC
	b	i	Thrombin	1	C
		ii	Fibrinogen	1	C
16	a	i	Pancreas	1	C
		ii	Glucagon	1	C
		iii	Glycogen	1	C
	b		Patients with type 2 diabetes still produce insulin/their cells are less sensitive to it	1	A
17	a	i	Medulla	1	C
		ii	Antagonistic	1	C
	b	i	Diverging	1	C
		ii	Impulses from one neuron travel to/affect several neurons so affecting more than one destination/muscle/effector at the same time	1	C
18	a		Send nervous impulses to the muscles **AND** glands	1	C
	b		Language processing/personality/imagination/intelligence **Any 1**	1	C
	c		Corpus callosum = 1 Transfers information between the cerebral hemispheres = 1	2	CC
19	a		Items become displaced because short-term memory (STM) has a limited span **OR** Items decay because STM has a limited duration/time it retains information	1	C
	b	i	Rehearsal **OR** organisation	1	C
		ii	Elaborative encoding involves additional/meaningful information (to provide context for memory)	1	C
	c		Working memory model	1	C
20	a	i	Code for the synthesis/production of the neurotransmitters	1	A
		ii	Less myelination/myelin **AND** so impulses are not transmitted as quickly/responses are not as rapid	1	A
		iii	Glial cells	1	C
	b		By degradation by enzymes **OR** By reabsorption/reuptake by pre-synaptic membrane	1	C

ANSWERS TO PRACTICE QUESTIONS

Question			Expected answer	Mark	Demand
21	a		Mast cell	1	C
	b		Vasodilation = 1 Increased permeability of capillaries = 1	2	CC
	c		Result in accumulation of phagocytes at infection sites	1	C
22	a	i	Antigens	1	C
		ii	B lymphocyte	1	C
		iii	The antibodies only recognise the antigen from the polio virus	1	A
	b		T lymphocytes	1	C
23	a	i	When a threshold/large percentage of a population is immunised (against an infection) protecting those who are not immunised	1	C
		ii	Type of disease/effectiveness of vaccine/density of population **Any 1**	1	C
	b	i	Dead pathogens/weakened pathogens/parts of pathogen/inactivated toxin from pathogen **Any 1**	1	C
		ii	Adjuvant is a substance which makes the vaccine more effective/ enhances the immune response	1	C
24	a		Scientifically planned experimental procedure to test that a new drug/vaccine/treatment is safe/effective for licensing	1	C
	b		Randomised trials/double-blind trials/placebo-controlled trials **Any 2**	2	CC
	c		To increase the reliability of the data obtained/reduce experimental error/to establish statistical significance of the results	1	C

Data-handling questions

Question			Expected answer	Mark	Demand
1	a		Production of CO_2 **OR** production of heat	1	C
	b	i	1:20	1	C
		ii	8000 cm³ **OR** 8.0 litres	1	A
	c		Increasing exercise from standing to jogging increased the pulse rate slowly = 1 Increasing to running and then sprinting increased the pulse rate quickly = 1	2	CA
	d	i	Sprinting	1	C
		ii	130 W (Accept answers in the range 120–140 W)	1	C
	e		400%	1	A
2	a	i	Height increases from 80 cm at 1 to 188 cm at 17 = 1 Between 17 and 21 the height remains constant = 1	2	CA
		ii	4.6 kg	1	A
		iii	23.77/23.8 **AND** Healthy	1	A
		iv	64.8 kg	1	A
	b	i	As BMI increases, the relative risk of developing diabetes increases	1	C
		ii	13:1	1	A
		iii	700%	1	A

ANSWERS TO PRACTICE QUESTIONS

Question			Expected answer	Mark	Demand
3	a	i	16 units	1	C
		ii	25%	1	A
		iii	From 0–5 nM increases from 0 to 20 units = 1 From 5–25 nM increases from 20 to 37 units = 1	2	CA
		iv	39 units	1	C
	b		6.0 units	1	A
	c		8.5 nM	1	A
	d		20%	1	A

Experimental questions

Question			Expected answer	Mark	Demand
1	a		Temperature **OR** type/concentration of stain **OR** same type/concentration of enzymes used	1	C
	b		Run more samples of patient's and donors' DNA	1	C
	c		32 mm	1	C
	d		Scales and labels with units = 1 Accurately plotted points joined with straight lines = 1	2	CA
	e		The larger the fragment/more bp it has, the shorter the distance it travels in the gel (or converse)	1	C
	f		900 base pairs	1	C
	g		Donor 3	1	C
2	a		5%	1	A
	b	i	Blood pressure after taking the energy drink	1	C
		ii	Other drinks/food taken (before or after) **OR** Concentration of energy drink **OR** Activities carried out during the hour	1	A
	c		Another similar group of volunteers who were treated in the same way but were not given an energy drink	1	C
	d		Scales and labels from table = 1 Stacked bars accurately plotted and with straight tops = 1	2	CA
	e	i	Energy drinks increase systolic/diastolic BP	1	C
		ii	Ages/genders/weights of participants not taken into account/small sample size	1	C
3	a	i	Scales and labels with units = 1 Accurately plotted points joined with straight lines = 1	2	CC
		ii	As the position of the word increases the percentage of correct responses decreases from 100% at position 1 to 40% at position 8 = 1 As the position of the word increases from 8 to 14 the percentage of correct responses increases to 100% and then remains constant = 1	2	CA
		iii	Results from 20 participants	1	C
		iv	8	1	C
	b	i	Decrease in the ability to recall the position of the word in the series at/towards the end of the list/at positions 11–15	1	C
		ii	Age of students/time of day/surroundings	1	C

Mini extended response questions

Question	Expected answer	Mark	Demand
1	1 Both types of stem cell can continue to divide 2 Both types of stem cell can differentiate into specialised cells 3 Embryonic stem cells are pluripotent/have the potential to become any cell type 4 Tissue stem cells are multipotent/give rise to a limited range of cell types 5 Tissue stem cells develop into cell types that are closely related to the tissue in which they are found/divide and differentiate to replenish cells that need to be replaced 6 Tissue stem cells in bone marrow give rise to various types of blood cell Any 5	5	CCCAA
2	1 DNA heated to 92–98 °C to separate strands/break hydrogen bonds between base pairs 2 Cooled to between 50–65 °C to allow primers to bind to target sequences of DNA 3 Primers are complementary to specific target sequences at the two ends of the region to be amplified/primers bind to the 3′ end of the DNA strands 4 Heat-tolerant DNA polymerase then replicates the (primed region of) DNA at 70–80 °C 5 Repeated cycles of heating and cooling amplify this region of DNA Any 4	4	CCCA
3	1 During exercise (muscles) do not have enough oxygen for (aerobic) respiration/the electron transport chain 2 Pyruvate is converted to lactate 3 Transfer of hydrogen ions and electrons from NADH (produced during glycolysis) 4 Regenerates NAD needed to maintain ATP production by glycolysis 5 Lactate builds up (in muscles) causing fatigue/an oxygen debt 6 Lactate is converted back into pyruvate Any 5	5	CCCAA
4	1 Ovulation is stimulated by drugs preventing negative feedback control 2 These drugs prevent the negative feedback effect of oestrogen on FSH secretion 3 FSH continues to be released and ova are produced 4 Some ovulatory drugs mimic FSH and LH 5 Drugs that mimic FSH result in follicular development 6 Drugs that mimic LH stimulate ovulation Any 5	5	CCCAA

ANSWERS TO PRACTICE QUESTIONS

Question	Expected answer	Mark	Demand
5	1 The medulla in the brain regulates the rate of the SAN through the antagonistic action of the autonomic nervous system (ANS) 2 Sympathetic/accelerator nerves release noradrenaline 3 Noradrenaline increases the heart rate 4 Parasympathetic nerves release acetylcholine 5 Acetylcholine decreases the heart rate Any 4	4	CCCA
6	1 The accumulation of fatty material/cholesterol (and fibrous material and calcium) forming an atheroma/plaque 2 An atheroma forms beneath the endothelium of the artery wall 3 Artery thickens/lumen diameter reduced 4 Artery loses its elasticity 5 Blood flow is restricted and results in increased blood pressure 6 Atherosclerosis is the cause of various cardiovascular diseases (CVD)/ angina/myocardial infarction (MI)/stroke/peripheral vascular disorders Any 5	5	CCCAA
7	1 Converging pathway has several neurons linked to one neuron 2 This increases the neurotransmitter concentration/chances of impulse generation/sensitivity to excitatory or inhibitory signals 3 Diverging pathway has one neuron linked to several neurons 4 This means that impulses are sent to/influence several destinations/ muscles/effectors at the same time 5 In reverberating pathways, neurons later in the pathway synapse/link with neurons earlier in the pathway Any 4	4	CCCA
8	1 Long-term memory (LTM) has an unlimited capacity 2 Items are transferred from STM by rehearsal 3 By organisation 4 By elaboration 5 Information can be encoded by shallow or elaborative means 6 Retrieval is aided by contextual cues Any 5	5	CCCAA
9	1 Immunisation can be given through vaccination 2 Vaccines are based on antigens that have been rendered harmless 3 Dead/weakened/inactivated/parts of antigens/toxins are used in vaccines 4 Adjuvants are added to enhance the immune response 5 Memory cells are produced during the immune response/ immunological memory created 6 Memory cells remain in the body for many years after the vaccination 7 Produce more rapid/greater immune response following next exposure to the antigen Any 5	5	CCCAA

ANSWERS TO PRACTICE QUESTIONS

Full extended response questions

Question		Expected answer	Mark	Demand
1		1 Double strand of nucleotides/double helix 2 Deoxyribose sugar, phosphate and base 3 Sugar phosphate backbone 4 Complementary base pairs of adenine–thymine and cytosine–guanine 5 Hydrogen bonds between bases 6 Antiparallel structure with deoxyribose and phosphate at 3′ and 5′ ends 7 DNA is associated with/tightly coiled around histones 8 DNA unwinds into two strands 9 Primers needed to start replication/attach to target sequence at 3′ end 10 DNA polymerase adds nucleotides to 3′ (deoxyribose) end of strand 11 DNA polymerase adds nucleotides in one direction 12 One strand/leading strand replicated continuously, the other/lagging strand is replicated in fragments 13 Fragments joined by ligase **Any 9**	9	CCCCCCAAA
2		1 Anabolic pathways require energy 2 Catabolic pathways release energy 3 Reversible and irreversible pathways 4 Regulation by intra- and extracellular molecules 5 Activation energy lowered by enzymes 6 Enzymes have affinity for substrate but less for product molecules 7 Induced fit of enzymes to substrates 8 Fit occurs at active site 9 Enzyme reaction rate can be reduced by inhibitors 10 Competitive inhibitors block active sites 11 Non-competitive inhibitors bind to other regions of enzyme molecule 12 Feedback/end-product inhibition **Any 9**	9	CCCCCCAAA
3	a	1 Pituitary gland secretes/produces FSH/LH 2 FSH stimulates growth of follicle (in the ovary) 3 Follicle/ovary produces oestrogen 4 Oestrogen stimulates/repairs the endometrium/uterus lining **OR** stimulates vascularisation of the endometrium 5 Oestrogen stimulates production/surge of LH 6 LH (surge) brings about ovulation/release of the egg 7 Rising/high levels of oestrogen inhibit FSH production 8 This is negative feedback **Any 5**	5	CCCCA

ANSWERS TO PRACTICE QUESTIONS

Question		Expected answer	Mark	Demand
	b	1 The follicle develops into the corpus luteum 2 Corpus luteum secretes progesterone 3 Progesterone maintains/increases/thickens the endometrium/uterus lining **OR** increases vascularisation 4 High levels of progesterone/oestrogen inhibit FSH/LH production 5 Corpus luteum degeneration results in a drop in progesterone level 6 Fall in progesterone level triggers menstruation/breakdown of the endometrium **Any 4**	4	CCCA
4	a	1 Pacemaker/SAN contains autorhythmic cells/is where the heartbeat originates/is found in the right atrium 2 Impulse/wave of excitation spreads across the atria/causes the atria to contract/causes atrial systole 3 (Impulses) reach/stimulate the atrioventricular node/AVN 4 AVN found at junction of atria and ventricles/at base of atria 5 Impulses from AVN spread through ventricles 6 (Cause) contraction of ventricles/ventricular systole 7 (This is followed by) relaxation/resting/diastolic phase/diastole **Any 4**	4	CCCA
	b	1 The autonomic nervous system/ANS controls the cardiac cycle/sinoatrial node/SAN 2 The ANS is located in the medulla 3 Sympathetic nerves speed up the heart rate 4 Sympathetic nerves release noradrenaline 5 Parasympathetic nerves slow down the heart rate 6 Parasympathetic nerves release acetylcholine 7 Sympathetic and parasympathetic systems are antagonistic to each other **Any 5**	5	CCCAA
5	a	1 Synapse/synaptic cleft is the junction/gap between neurons/nerve cells 2 Neurotransmitters are stored in/released from vesicles on arrival of impulse 3 Neurotransmitters diffuse across the gap 4 Neurotransmitters bind with/reach receptors 5 Threshold/minimum number of neurotransmitters is needed (for the impulse to continue) 6 Neurotransmitters are removed by enzymes **OR** reuptake/reabsorption 7 Neurotransmitters must be removed to prevent continuous stimulation 8 Named neurotransmitter – acetylcholine/noradrenaline/dopamine/endorphins **Any 5**	5	CCCCA

Question		Expected answer	Mark	Demand
	b	1 Converging pathway has several neurons linked to one neurone 2 This increases the neurotransmitter concentration/chances of impulse 3 Generation/sensitivity to excitatory or inhibitory signal 4 Diverging pathway has one neuron linked to several neurons 5 This means that impulses are sent to/influence several destinations/effectors at the same time 6 Reverberating pathways – neurons later in pathway synapse/link with neurons earlier in the pathway **Any 4**	4	CCCA
6	a	1 Immunisation can be given through vaccination 2 Vaccines are based on antigens which have been rendered harmless 3 Dead/weakened/inactivated/parts of antigens/toxins are used 4 Adjuvants added to enhance the immune response 5 Memory cells produced during the immune response/ immunological memory created 6 Remain in the body for many years after the vaccination 7 Produce more rapid/greater immune response following exposure to antigen **Any 5**	5	CCCAA
	b	1 Large number of individuals in a population are immunised 2 Non-immune individuals are protected 3 There is a lower probability that they will come into contact with an infected individual 4 Threshold percentage of population varies with the type of disease/the effectiveness of the vaccine/the density of the population 5 Poverty prevents widespread vaccination 6 Parental resistance to allowing their children to be vaccinated 7 Malnutrition may make vaccination unsafe **Any 4**	4	CCCA

PRACTICE EXAM A

Paper 1

Total marks: 25

Attempt ALL questions.

The answer to each question is either A, B, C or D. Decide what your answer is, then circle the appropriate letter.

There is only one correct answer to each question.

Allow yourself 40 minutes for Paper 1.

STUDENT MARGIN

1. Which of the following diagrams shows how strands of a DNA molecule are arranged?

   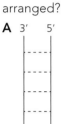 

   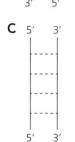 

   Applying KU

2. If 10% of the bases in a molecule of DNA are cytosine, what is the ratio of cytosine to thymine in the same molecule?
   - A 1 : 1
   - B 1 : 2
   - C 1 : 3
   - D 1 : 4

   Applying KU

3. Which pathway describes the production of haploid gametes from diploid germline cells?
   - A diploid germline cell → mitosis → diploid germline cell → meiosis → haploid gamete
   - B diploid germline cell → mitosis → diploid germline cell → mitosis → haploid gamete
   - C diploid germline cell → meiosis → diploid germline cell → meiosis → haploid gamete
   - D diploid germline cell → meiosis → diploid germline cell → mitosis → haploid gamete

   Demonstrating KU

52

PRACTICE EXAM A

STUDENT MARGIN

4 The diagram represents a stage in protein synthesis in a cell.

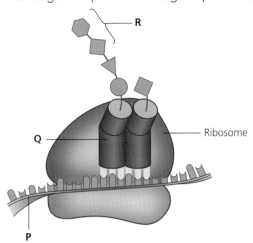

Which row in the table identifies molecules P, Q and R?

	Molecules		
	P	Q	R
A	tRNA	mRNA	peptide
B	mRNA	peptide	tRNA
C	tRNA	peptide	mRNA
D	mRNA	tRNA	peptide

Applying KU

5 The graph shows the temperature changes involved in one thermal cycle of the polymerase chain reaction (PCR).

Which letter on the graph indicates a region in which DNA polymerase would be most active?

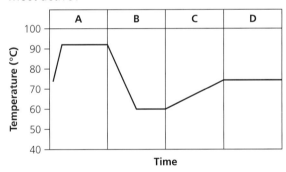

Applying KU

6 The graph shows the effect of substrate concentration on the rate of an enzyme-catalysed reaction.

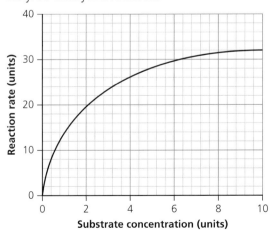

53

At which substrate concentration is the reaction rate equal to 75% of the maximum rate?

A   0.4 units
B   2.6 units
C   3.2 units
D   6.0 units

7   In a metabolic pathway, feedback inhibition can occur when its
A   end product binds to a substrate involved in an earlier step in the pathway
B   final enzyme binds to a substrate involved in an earlier step in the pathway
C   end product binds to an enzyme involved in an earlier step in the pathway
D   final enzyme binds to a product of an earlier step in the pathway.

8   The graph shows the energy changes involved in a chemical reaction.
Which letter indicates the activation energy of the reaction in the absence of an enzyme specific to this substrate?

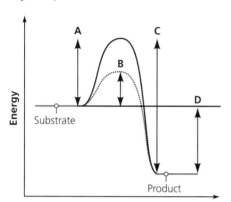

9   Alcaptonuria is a metabolic disorder caused by a failure to produce the enzyme that catalyses the breakdown of homogentisate into maleylacetoacetate, as shown in the metabolic pathway below.

Amino acids from the diet
↓
Hydroxyphenylpyruvate (HPP)
↓
Homogentisate → Benzoquinone acetic acid (BQA)
                  (which causes the symptoms of alcaptonuria)
Metabolic block ⊘
↓
Maleylacetoacetate

From the information given, which row in the table shows the likely effect of this metabolic block on the concentration of the substances shown?

	Increased concentration	Decreased concentration
A	HPP	maleylacetoacetate
B	BQA	homogentisate
C	BQA	maleylacetoacetate
D	HPP	homogentisate

10 The diagram shows a section through the seminiferous tubules in the testes. Which letter indicates a target cell for the hormone ICSH?

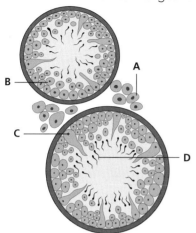

11 As part of an investigation into human fertility, mean sperm counts were calculated from semen samples of groups of men over the period between 1940 and 2000.

The results are shown on the graph.

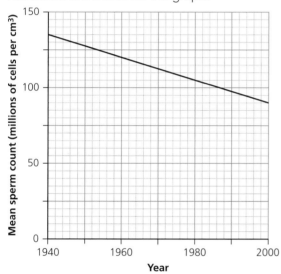

What is the average reduction in mean sperm count per year over the period of the study?

A 0.67 million per cm³
B 0.75 million per cm³
C 0.92 million per cm³
D 45.0 million per cm³

12 Which of the following diagrams best represents a cross-section through an artery?

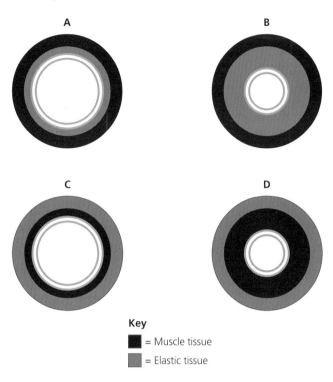

Key
■ = Muscle tissue
▨ = Elastic tissue

13 The diagram shows the relationship between blood capillaries, body cells and lymph capillaries.

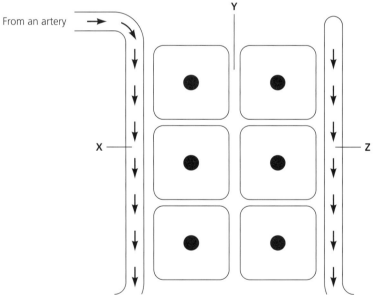

In which region(s) of the diagram would plasma protein molecules be found?

A   X only
B   X and Y
C   Y and Z
D   Z only

14. Mean arterial pressure (MAP) is a measure of blood pressure in arteries.
Pulse pressure is the difference between systolic and diastolic blood pressure.
MAP is calculated using the formula below.

MAP = diastolic pressure + $\frac{\text{pulse pressure}}{3}$

What is the MAP of an individual with a blood pressure reading of 120/75 mmHg?

A   90 mmHg
B   120 mmHg
C   135 mmHg
D   165 mmHg

15. Which row in the table describes how the ratio of lipoproteins affects the level of cholesterol in blood and the risk of atherosclerosis?

	Ratio of HDL : LDL	Effect on blood cholesterol	Risk of atherosclerosis
A	lower	lower	increased
B	higher	higher	increased
C	higher	lower	reduced
D	lower	higher	reduced

16. The graph shows how infant mortality in a country without a vaccination programme changed over a 7-year period. Also shown is the change in the percentage of mothers breastfeeding their babies.

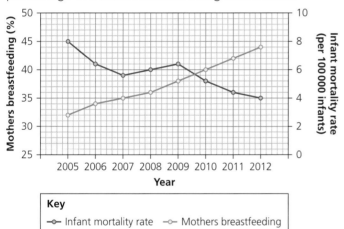

What percentage of babies died when 40% of mothers were breastfeeding?

A   0.0046
B   0.0052
C   4.600
D   5.200

PRACTICE EXAM A

17 Which row in the table describes the effects of oestrogen in fertility?

	Effect on endometrial proliferation	Effect on cervical mucus consistency
A	stimulated	increased chance of sperm penetration
B	inhibited	increased chance of sperm penetration
C	stimulated	decreased chance of sperm penetration
D	inhibited	decreased chance of sperm penetration

18 PKU is caused by a

  A substitution mutation, making the enzyme that converts phenylalanine to tyrosine non-functional

  B deletion mutation, making the enzyme that converts phenylalanine to tyrosine non-functional

  C substitution mutation, making the enzyme that converts tyrosine to phenylalanine non-functional

  D deletion mutation making the enzyme that converts tyrosine to phenylalanine non-functional.

19 Which row in the table identifies actions of branches in the autonomic nervous system?

	Parasympathetic	Sympathetic
A	heart rate increased	heart rate decreased
B	increased release of intestinal secretions	decreased release of intestinal secretions
C	decreased rate of peristalsis	increased rate of peristalsis
D	increased breathing rate	decreased breathing rate

20 Heroin can act as an agonist drug and its use can lead to drug tolerance in which the user must take more of the drug to get an effect.

The reason for tolerance is that the heroin causes

  A a decrease in the number of its receptors and desensitisation

  B an increase in the number of its receptors and desensitisation

  C a decrease in the number of its receptors and sensitisation

  D an increase in the number of its receptors and sensitisation.

21 The diagram shows features of the relationship between short- and long-term memory.

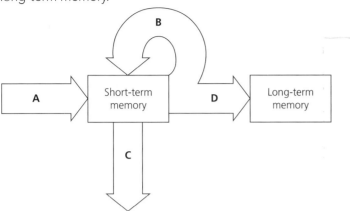

Which arrow could represent the process of displacement?

22 Acquired immune deficiency syndrome (AIDS) is a condition that may develop from an HIV infection. The graph shows the number of people in the world population infected by HIV and the number of deaths from AIDS between 1990 and 2010.

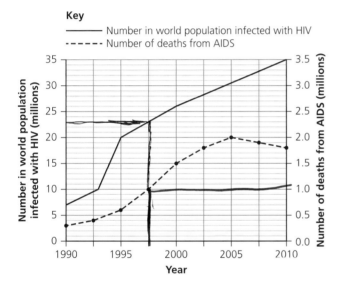

During the year in which 1 million people died from AIDS, how many millions of people were infected with HIV?

A 1.8
B 10
C 23
D 26

23 Adjuvants are often added to vaccines to

A make the vaccines safer
B enhance the immune response that the vaccines trigger
C make the immunity that the vaccines produce last longer
D ensure total removal of pathogens from the vaccines.

24 The graphs show the effects of two injections of an antigen on the concentrations of the antibody produced against it in the blood of a patient.

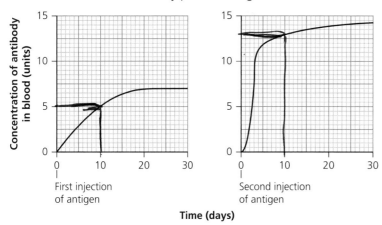

Time (days)

What is the percentage increase in the concentration of antibodies in the blood of the patient 10 days after the second injection compared to 10 days after the first?

A 38.0
B 62.5
C 160
D 260

25 Which of the following is the principle of herd immunity?

A Non-immune individuals are protected as there is a lower probability they will come into contact with uninfected individuals.
B Non-immune individuals are protected as there is a lower probability they will come into contact with infected individuals.
C Immune individuals are protected as there is a lower probability they will come into contact with infected individuals.
D Immune individuals are protected as there is a lower probability they will come into contact with uninfected individuals.

[End of Paper 1]

Paper 2

Total marks: 95

Attempt ALL questions.

Question 19 contains a choice.

Write your answers clearly in the spaces provided. If you need additional space for answers or rough work, please use separate pieces of paper.

Allow yourself 2 hours and 20 minutes for Paper 2.

1 The diagram shows the role of embryonic stem cells in the development of a human embryo.

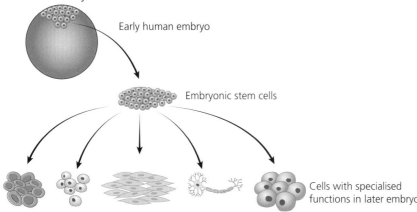

a Give the term used to describe the process by which a cell develops specialised functions. — 1 mark (Demonstrating KU)

b Describe the difference in function between embryonic stems cells and tissue stem cells. — 2 marks (Applying KU)

c Describe how cancer cells form a tumour and explain how secondary tumours can arise. — 2 marks (Demonstrating KU)

Description

Explanation

PRACTICE EXAM A

2 The diagram shows part of a DNA template strand and part of a primary RNA transcript synthesised from it.

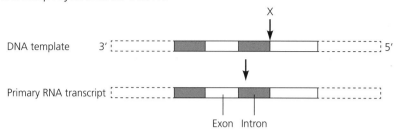

a Give the term used to describe the process shown in the diagram. **[1]**

b DNA is encoded in triplet sequences.
Explain what is meant by this. **[1]**

c Name the enzyme responsible for synthesising the primary RNA transcript. **[1]**

d Describe a possible effect on the primary RNA transcript of a single nucleotide mutation at point X on the DNA template. **[1]**

3 a The table shows single nucleotide substitution mutations of human genes and the possible effect they may have.
Complete the table by adding correct information to the empty boxes. **[2]**

Name of single nucleotide substitution	Possible effect of the mutation on the protein synthesised
	A correct amino acid replaced by an incorrect one in a polypeptide chain
Nonsense	

b One form of Down syndrome is caused by a translocation mutation that produces substantial changes to an affected individual's genetic material.
  i Describe what is meant by translocation. **[1]**

  ii Apart from translocation, name **one other** type of mutation that can affect the structure of human chromosomes. **[1]**

PRACTICE EXAM A

4 The diagram represents molecules involved in an enzyme-catalysed reaction in the presence of an inhibitor.

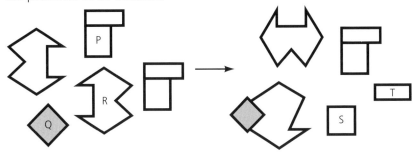

a Use letters from the diagram to identify an inhibitor molecule and a substrate molecule.

Molecule	Letter from diagram
inhibitor	
substrate	

2 — Demonstrating KU

b Name the type of inhibition occurring in this example and explain how the inhibitor molecules produce their effect.

Type of inhibition

_____

Explanation

_____

_____

2 — Applying KU

5 During strenuous exercise, the following processes occur in muscle cells.
The substance creatine phosphate, which is a reserve of phosphate, is broken down to release energy and phosphate that are used to produce ATP, as shown in the diagram.
Pyruvate is converted to lactate as oxygen becomes deficient.
The graph shows the concentrations of creatine phosphate and lactate in the muscle cells of a middle-distance runner over a 20-second period on a treadmill, during which he jogged gently for the first 10 seconds then sprinted strenuously for 10 seconds.

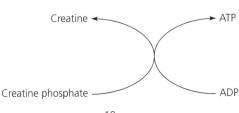

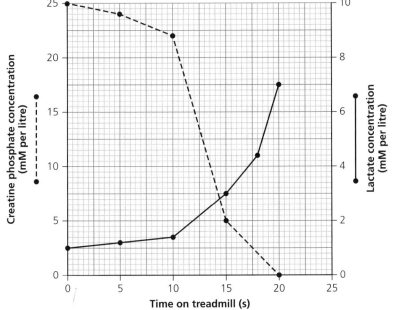

PRACTICE EXAM A

a **i** Give the lactate concentration in muscle cells after 15 seconds.

_____ mM per litre

[1 mark — Selecting]

  ii Calculate the average increase in lactate concentration per second over the total 20 seconds.

Space for calculation

_____ mM per litre

[1 mark — Processing]

  iii Give the creatine phosphate concentration when the lactate concentration was 5 mM per litre.

_____ mM per litre

[1 mark — Selecting]

b Using the information given, explain the reasons for the changes in concentration of the two substances shown in the graph.

Creatine phosphate

_____

_____

Lactate

_____

_____

[2 marks — Applying KU]

c The chart shows the percentages of fast- and slow-twitch muscle fibres in the muscles of the middle-distance runner, compared with those of athletes in other categories and in untrained individuals.

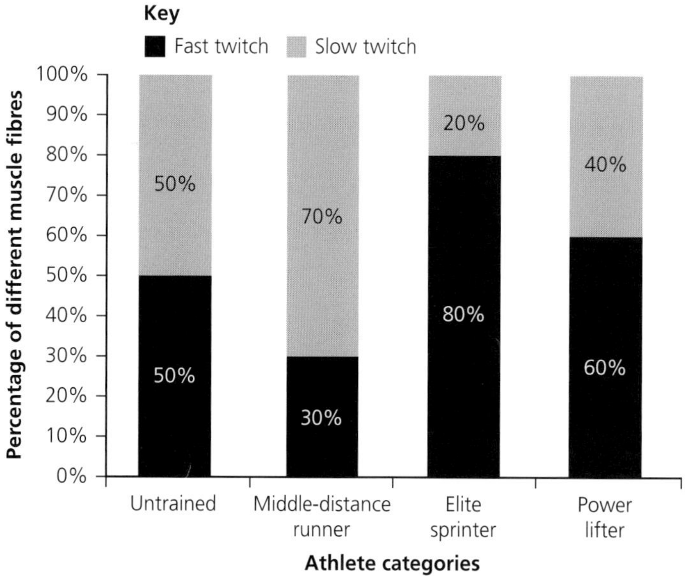

Key
■ Fast twitch   ■ Slow twitch

Percentage of different muscle fibres:
- Untrained: Fast twitch 50%, Slow twitch 50%
- Middle-distance runner: Fast twitch 30%, Slow twitch 70%
- Elite sprinter: Fast twitch 80%, Slow twitch 20%
- Power lifter: Fast twitch 60%, Slow twitch 40%

Athlete categories

  i Describe the differences in the percentages of muscle types in the middle-distance runner compared with:

an untrained individual

_____

_____

[2 marks — Selecting]

MARKS | STUDENT MARGIN

an elite sprinter.

____

____

ii Explain how the percentages of the different fibres found in the muscles of power lifters are suitable for their activity. **1** Applying KU

____

____

6 The diagram shows stages in the aerobic respiration of glucose in a cell.

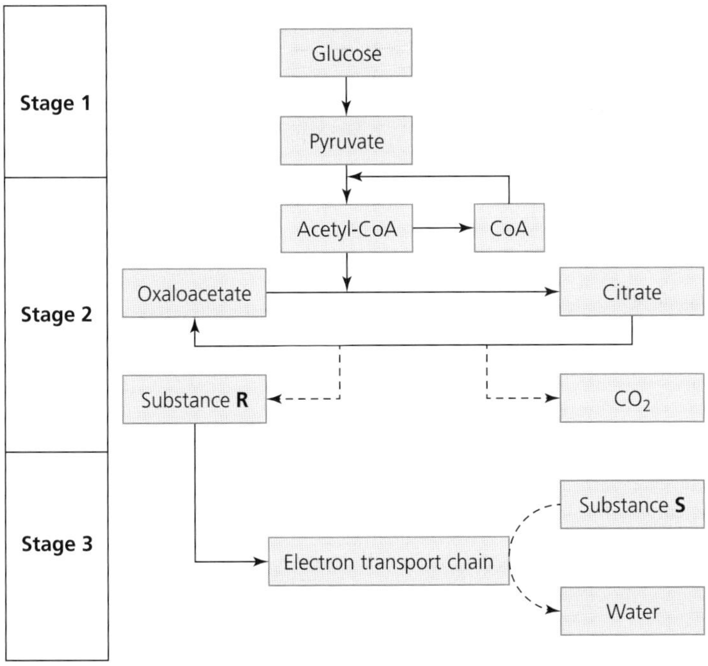
Stage 1
Stage 2
Stage 3
Glucose
Pyruvate
Acetyl-CoA
CoA
Oxaloacetate
Citrate
Substance **R**
CO_2
Substance **S**
Electron transport chain
Water

a Name stage 1. **1** Demonstrating KU

____

b Name substance R. **1** Applying KU

____

c Identify substance S and describe its role in stage 3. **2** Applying KU

Substance S

____

Role in stage 3

____

____

7 The graph shows the relative concentrations of three hormones in the blood plasma of a woman during a 28-day menstrual cycle.

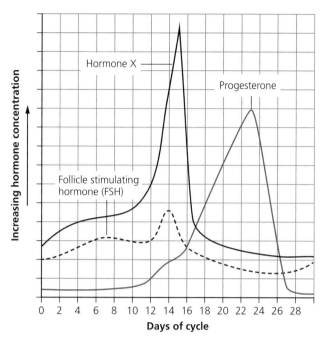

a Name hormone X, which triggers ovulation.

b Describe the effects of progesterone on the uterus as it:
   i increases in concentration from about day 12

   ii decreases in concentration after day 23.

c Name the gland that releases follicle stimulating hormone (FSH) and describe the role of FSH in fertility.

   Name

   Role in fertility

MARKS | STUDENT MARGIN

8 The family history chart shows the inheritance of haemophilia in a family.

The allele for haemophilia (h) is sex-linked and recessive to the normal allele (H).

Key

- ☐ Male without the condition
- ■ Male with the condition
- ○ Female without the condition
- ● Female with the condition

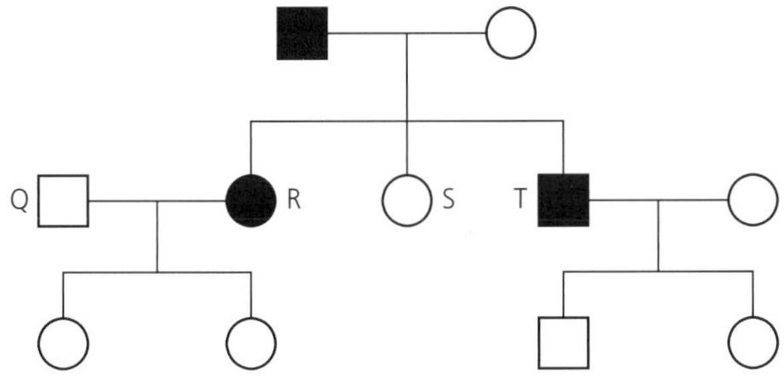

a Explain why individual R has haemophilia even though her mother was not affected. **1** Applying KU

____

____

b Give the genotype of individual S. **1** Applying KU

____

____

c Individuals Q and R are expecting their third child.

Following the results of an amniocentesis test, the parents are told that their expected baby will be male.

i Describe what is meant by an amniocentesis test and explain how the test can reveal the gender of an unborn baby. **2** Demonstrating KU

Meaning

____

____

Explanation

____

____

ii Calculate the percentage chance that the expected male baby of individuals Q and R will have haemophilia. **1** Processing

Space for calculation

__________ %

9 Gastric band surgery (GBS) can be used to treat individuals with obesity. In a clinical trial, sensitivity to insulin was measured in groups of GBS patients with and without type 2 diabetes. In type 2 diabetes, liver cell sensitivity to insulin is low. The patients' sensitivity to insulin was measured before and after the GBS procedure. The mean results, as well as the ranges of values obtained in the trial, are shown in the table.

The higher the number of units, the greater the sensitivity to insulin.

MARKS STUDENT MARGIN

Patient group	Mean insulin sensitivity (units)	
	Before GBS procedure	One month after GBS procedure
non-diabetic	0.55 (± 0.32)	1.30 (± 0.88)
type 2 diabetes	0.40 (± 0.24)	1.10 (± 0.87)

a Use the data in the table to support the following conclusions.

i Non-diabetic patients were at higher risk of developing diabetes following the GBS procedure. **1** Selecting

______

______

______

ii GBS helped many patients with type 2 diabetes but some were not helped. **2** Selecting

______

______

______

b Describe the effects of insulin on liver cells in non-diabetic individuals. **2** Applying KU

______

______

______

10 The diagram shows stages in thrombosis within a blood vessel.

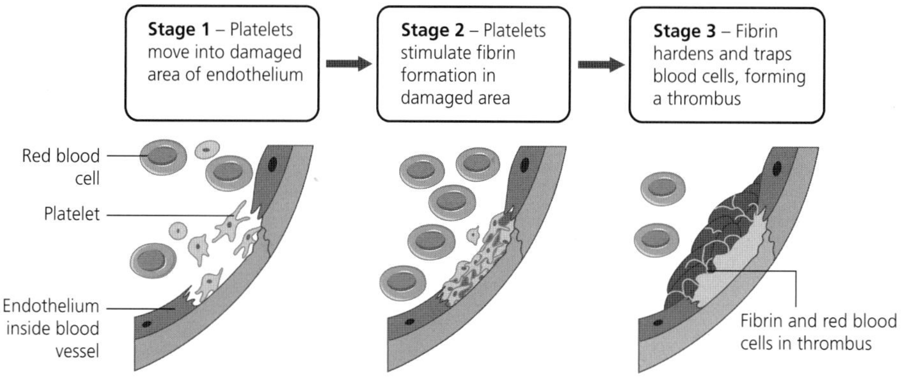

a Suggest how damage to the endothelium can occur. **1** Demonstrating KU

______

______

b Name the soluble protein present in blood plasma from which fibrin is produced. **1** Demonstrating KU

______

c Describe how thrombosis can lead to myocardial infarction (MI). **2** Applying KU

______

______

______

11 The diagram shows the heart and some of its associated nerves.

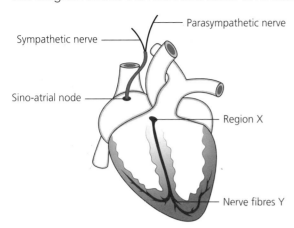

a Name the region of the brain that regulates the sinoatrial node (SAN) through the action of the sympathetic and parasympathetic nerves shown.

b The sympathetic and parasympathetic nerves act antagonistically. Explain the meaning of this statement with reference to heart rate.

c Name region X.

d Describe the role of region X and nerve fibres Y in the cardiac cycle.

12 The diagram represents the structure of the heart and its associated blood vessels. The four chambers of the heart are labelled P–S.

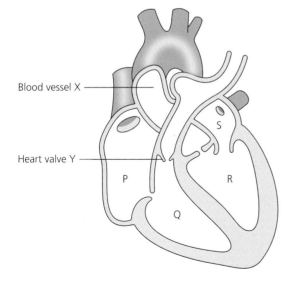

a Name blood vessel X and heart valve Y.

Blood vessel X _____ Heart valve Y _____

PRACTICE EXAM A

b i Using the letters to identify the heart chambers, explain what is meant by the term stroke volume (SV).

_____

ii Describe how the stroke volume can be used to calculate cardiac output (CO).

_____

13 Write notes on antenatal screening.

_____
_____
_____
_____
_____

14 The diagram shows some cells in the nervous system.

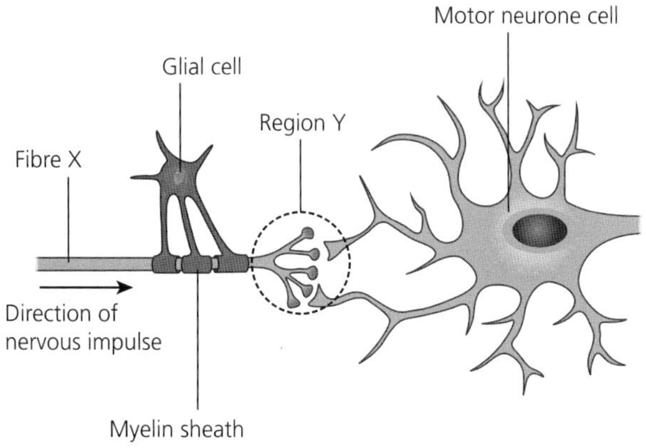

Labels: Motor neurone cell, Glial cell, Region Y, Fibre X, Direction of nervous impulse, Myelin sheath

a Name fibre X and give the function of the myelin sheath that has developed around it.

Fibre X
_____

Function of myelin sheath
_____

b Explain why the glial cell is able to synthesise the protein myelin, whereas other cells cannot.

_____

c In region Y, nervous impulses pass across synapses.

Describe how neurotransmitters are involved in the passage of nervous impulses across the synapse.

_____
_____

Marks	Student Margin
1	Applying KU
1	Demonstrating KU
4	Demonstrating KU
2	Applying KU
1	Applying KU
2	Demonstrating KU

15 In an experiment into the effects of caffeine on learning, thirty 25-year-old volunteers were split into three groups of ten individuals. Members of each group were given different dosages of caffeine as shown in **Table 1**.

Each individual was blindfolded and asked to try a finger maze, as shown in the diagram.

The number of errors made during each trial was recorded. Each individual completed six trials consecutively with no breaks between trials.

The results are shown in **Table 2**.

Table 1

Group	Caffeine dosage given (mg)
1	50
2	100
3	150

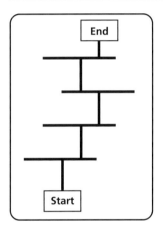

Table 2

Trial	Average number of errors per group		
	Group 1	Group 2	Group 3
1	7	6	7
2	6	6	5
3	4	3	2
4	2	2	1
5	1	0	0
6	0	0	0

a Give a hypothesis that could be tested by this experiment.

b Identify the dependent variable in this experiment.

c Identify **one** variable, other than those already mentioned, which would have to be kept constant to ensure that valid conclusions could be drawn from the results.

d Describe a suitable control for this experiment.

PRACTICE EXAM A

e On the grid below, plot line graphs to show all of the results of this experiment.

MARKS: 2 — Presenting

f Give a conclusion that could be drawn from the results of this experiment.

MARKS: 1 — Concluding

g Suggest an improvement to the experimental method that could increase the reliability of the results.

MARKS: 1 — Evaluating

16 In a study of brain activity, a description of a method to make a paper boat was read out to an individual. They were then asked to build the boat.
The levels of activity in different areas of part of the individual's brain were measured. The results are shown in the diagram.

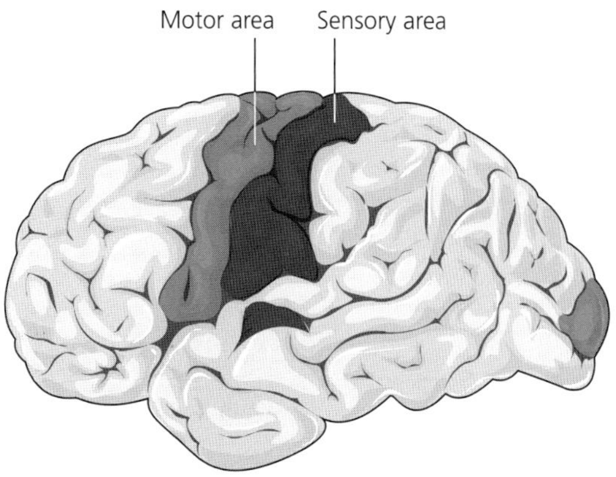

Motor area   Sensory area

■ Areas of high activity during the description of the task
■ Areas of high activity during the completion of the task

a Name the part of the brain shown in the diagram. [1]

b Describe the evidence in the diagram that confirms that there is localisation of function in the brain. [1]

c i Explain the high levels of activity in the motor and sensory areas of the brain during the completion of the task. [2]

ii Apart from the motor and sensory areas, name **one** other area found in the brain. [1]

17 During the planning of clinical trials to test a new drug, volunteers were recruited who were to be divided into two groups. One group was to receive the new drug and the other was to act as a control and so would not receive the drug.

a Describe how valid comparisons between the two groups can be achieved. [2]

b Suggest how the planners could ensure that a high degree of statistical significance in the results would be obtained. [1]

18 The bacterial species *Campylobacter*, *Salmonella* and *E. coli* 0157 each cause infections of the digestive system that result in vomiting and diarrhoea.
The chart shows the reported number of cases of these infections in a Scottish health board area between 1991 and 1996.

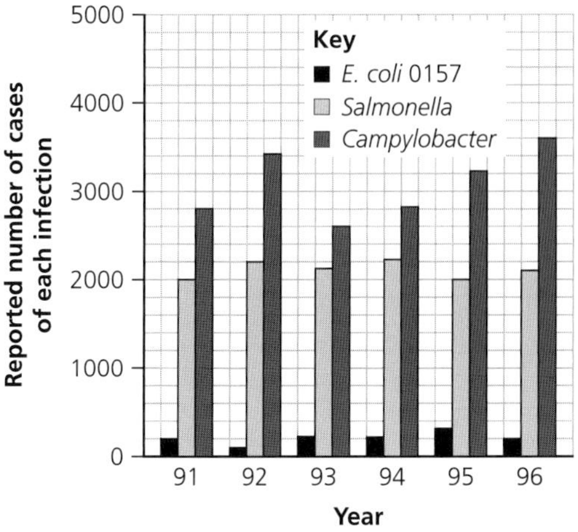

Key
- *E. coli* 0157
- *Salmonella*
- *Campylobacter*

a Use values from the chart to describe the changes in the number of reported cases of *Campylobacter* infections between 1991 and 1996. [2]

PRACTICE EXAM A

b Calculate the percentage increase in the **total** number of reported cases of these infections between 1991 and 1996.

Space for calculation

_____ %

c Suggest **one** precaution that could be taken to reduce the number of cases of these infections.

_____
_____

d Assuming that additional precautions were not taken, predict the number of cases of *Campylobacter* infections that could be expected in this health board area in 1997.

_____ cases

19 Answer **either question A or B** in the space below.

Labelled diagrams may be used where appropriate.

A Describe non-specific defences against disease.

OR

B Describe the roles of T and B lymphocytes in the specific immune response to disease.

_____
_____
_____
_____
_____
_____
_____
_____
_____
_____
_____
_____
_____
_____
_____
_____
_____
_____
_____
_____
_____

[End of Practice Exam A]

MARKS	STUDENT MARGIN
1	Processing
1	Planning
1	Predicting
8	Demonstrating KU

PRACTICE EXAM B

Paper 1

Total marks: 25

Attempt ALL questions.

The answer to each question is either A, B, C or D. Decide what your answer is, then circle the appropriate letter.

There is only one correct answer to each question.

Allow yourself 40 minutes for Paper 1.

STUDENT MARGIN

Demonstrating KU

1 Stem cells in bone marrow give rise to
   A platelets only
   B red blood cells only
   C red blood cells and platelets
   D red blood cells, platelets and phagocytes.

2 The graph shows changes in the number of human stem cells in a culture. The activity of the enzyme glutaminase, present in the cells, over an eight-day period is also shown.

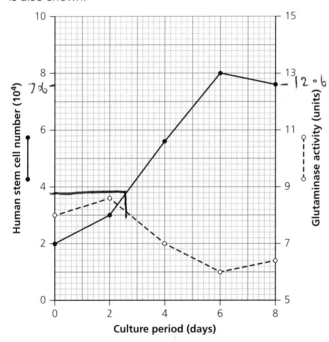

How many units of glutaminase activity were recorded when the cell number was 50% of its maximum over the eight days?

A 6
B 8
C 9
D 13

Selecting

3 The diagram represents a single molecule of transfer RNA (tRNA).

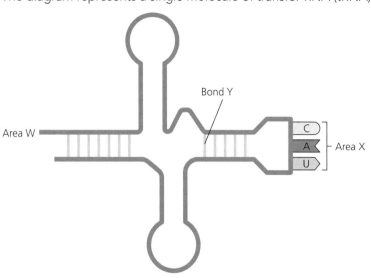

Which row in the table identifies the labelled parts of this molecule?

	Area W	Area X	Bond Y
A	anticodon	amino acid bonding site	peptide
B	amino acid bonding site	anticodon	hydrogen
C	anticodon	amino acid bonding site	hydrogen
D	amino acid bonding site	anticodon	peptide

4 The diagram represents the transcription and splicing stages in the expression of a gene.

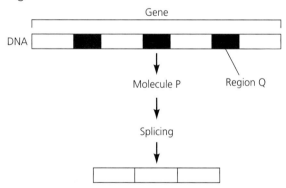

Which row in the table correctly identifies molecule P and region Q?

	Molecule P	Region Q
A	primary transcript	exon
B	messenger RNA	intron
C	messenger RNA	exon
D	primary transcript	intron

5. The diagram shows an enzyme molecule (E) bound to its substrate (S) and a molecule (I) that can act as an inhibitor of the enzyme.

Which row in the table identifies the type of inhibitor shown and the effect on its action of an increase in substrate concentration?

	Type of inhibitor	Effect of an increase in substrate concentration
A	competitive	reduction of inhibition
B	non-competitive	reduction of inhibition
C	competitive	no effect on inhibition
D	non-competitive	no effect on inhibition

6. Which of the following gene mutations would have a frameshift effect on the protein produced by its expression?
   A missense
   B nonsense
   C deletion
   D substitution

7. The use of genome information to identify the most appropriate drugs to treat illnesses is known as
   A personalised medicine
   B pharmacogenetics
   C bioinformatics
   D gel electrophoresis.

8. Which row in the table describes slow-twitch muscle fibres?

	Main energy storage substance	Relative number of mitochondria compared to fast-twitch fibres
A	fat	fewer
B	fat	more
C	glycogen	fewer
D	glycogen	more

PRACTICE EXAM B

9 The graph shows the concentration of lactate in the blood of an athlete and an untrained person during a 20-second period running on a treadmill.

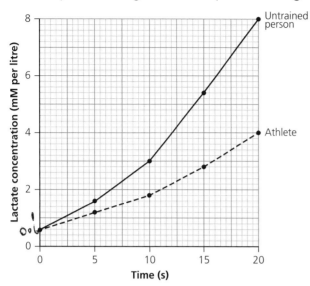

What was the average increase in lactate concentration per second in the blood of the untrained person over the period?

- A 0.17 mM per litre
- B 0.20 mM per litre
- C 0.37 mM per litre
- D 0.40 mM per litre

10 Cardiac output is calculated using the formula below.

cardiac output (l per minute) = heart rate (beats per minute) × stroke volume (cm^3)

The table shows the cardiac outputs and heart rates of four individuals.
Which individual has the greatest stroke volume?

	Cardiac output (l per minute)	Heart rate (beats per minute)
A	5.8	60
B	6.1	68
C	7.2	72
D	7.6	78

11 The diagram shows an external view of the human heart.

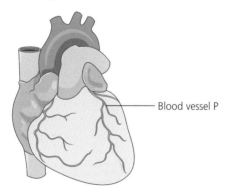

Identify blood vessel P.

- A aorta
- B vena cava
- C pulmonary artery
- D coronary artery

12 The ratio of high-density to low-density lipoproteins (HDL : LDL) in the blood is related to the level of cholesterol in the blood. Cholesterol level is related to chances of an individual developing cardiovascular disease (CVD).
Which row in the table below identifies these relationships?

	HDL : LDL	Level of cholesterol in the blood	Chances of developing CVD
A	high	low	decreased
B	high	high	increased
C	low	low	increased
D	low	high	decreased

13 The graph shows how the concentration of insulin in the blood of an individual was affected by changes in the concentration of glucose in their blood.
The individual has 4.8 litres of blood in their bloodstream.

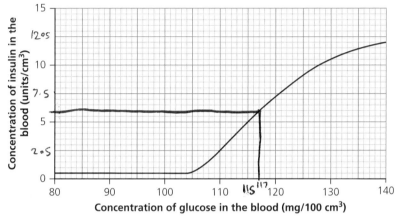

What is the total mass of glucose present in the bloodstream of the individual when their blood insulin concentration is 6 units?

A 114 mg
B 5472 mg
C 5616 mg
D 561 600 mg

14 Which row in the table identifies the cause and effect of type 1 diabetes?

	Cause	Effect
A	lack of insulin	failure to convert glucose to glycogen
B	cells lack sensitivity to insulin	failure to convert glycogen to glucose
C	lack of insulin	failure to convert glycogen to glucose
D	cells lack sensitivity to insulin	failure to convert glucose to glycogen

PRACTICE EXAM B

15 The family history chart shows the inheritance of Tay-Sachs disease in part of a family.

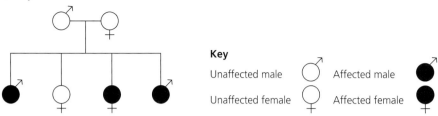

The information in the chart suggests that Tay-Sachs disease is caused by an allele that is

A   recessive and autosomal

B   dominant and autosomal

C   recessive and sex-linked

D   dominant and sex-linked.

16 The image shows a display of matched chromosomes.

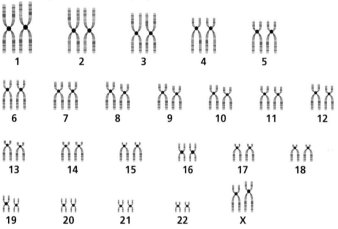

During antenatal care, which **two** techniques can be used to obtain cells for the production of an image such as this?

A   chorionic villus sampling (CVS) and amniocentesis

B   ultrasound imaging and chorionic villus sampling (CVS)

C   amniocentesis and pre-implantation genetic diagnosis (PGD)

D   pre-implantation genetic diagnosis (PGD) and ultrasound imaging

17 Which row in the table matches a neurotransmitter and information related to its functions?

	Neurotransmitter	Function 1	Function 2
A	dopamine	reduces feeling of pain	induces feeling of pleasure
B	endorphins	reduce feeling of pain	trigger release of sex hormones
C	dopamine	triggers release of sex hormones	reinforces behaviour in the reward pathway
D	endorphins	induce feeling of pleasure	reinforce behaviour in the reward pathway

18 A volunteer performed ten trials of a task involving mirror drawing. The graph shows the time taken to complete each trial and the number of errors made in each trial.

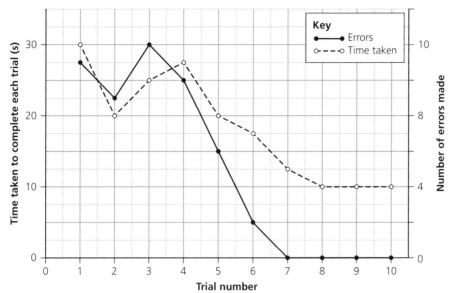

Which of the following conclusions is valid?

A Four errors were made in Trial 8.
B Trial 3 took 10 seconds to complete.
C No improvement in performance was noted after Trial 7.
D Overall performance improved between Trial 1 and Trial 4.

19 The table shows the relative number of deaths from various causes in the population of a developing country.

Cause of death	Number (millions)
infections and parasitic diseases	8.0
cancers	1.8
respiratory diseases	1.6
circulatory diseases	5.0
birth-related causes	2.0
other causes	1.6

What percentage of deaths was **not** due to infections and parasitic diseases?

A 12%
B 20%
C 40%
D 60%

20 Which of the following causes the production of antibodies in autoimmune disorders?

A viral infection
B self-antigens
C vaccination
D harmless antigens

PRACTICE EXAM B

21 Which row in the table shows the action of the autonomic nervous system?

	Speeding up breathing rate	Slowing down peristalsis	Speeding up production of intestinal secretions
A	sympathetic	sympathetic	parasympathetic
B	sympathetic	parasympathetic	parasympathetic
C	parasympathetic	sympathetic	sympathetic
D	parasympathetic	parasympathetic	sympathetic

22 The flowchart shows stages in the flow of information into long-term memory.

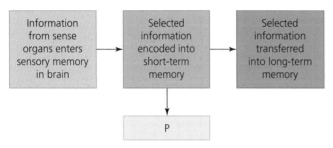

Which process is represented by box P?

A transfer to working memory
B shallow encoding
C displacement
D elaboration

23 Which type of protein molecules act as signals to specific white blood cells, causing them to accumulate at the site of an infection?

A hormones
B cytokines
C antibodies
D histamines

24 Which row in the table identifies the lymphocytes involved in allergy and describes their role in the condition?

	Lymphocytes involved	Role in the condition
A	B	attack body's own cells
B	B	respond to harmless antigens
C	T	attack body's own cells
D	T	respond to harmless antigens

25 The chart shows the results of a clinical trial of two influenza vaccines, K and M.

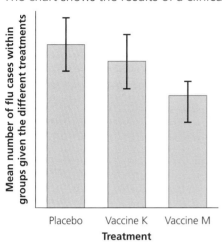

It can be concluded that

A vaccine K is more effective than vaccine M because there are fewer cases of flu than with the placebo treatment and error bars overlap with the placebo

B vaccine K is more effective than vaccine M because there are fewer cases of flu than with the placebo and the error bars overlap with vaccine M

C vaccine M is more effective than vaccine K because there are fewer cases of flu than with vaccine K and the error bars overlap

D vaccine M is more effective than vaccine K because there are fewer cases of flu than with the placebo and the error bars do not overlap with the placebo.

[End of Paper 1]

Concluding

Paper 2

Total marks: 95

Attempt ALL questions.

Question 17 contains a choice.

Write your answers clearly in the spaces provided. If you need additional space for answers or rough work, please use separate pieces of paper.

Allow yourself 2 hours and 20 minutes for Paper 2.

1. The diagram shows part of a DNA molecule at a stage in replication.

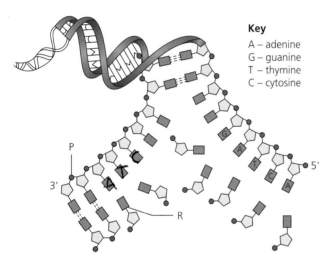

Key
A – adenine
G – guanine
T – thymine
C – cytosine

a Name molecule P, which forms parts of the backbone of a DNA strand. [1] *Applying KU*

b Name base R. [1] *Applying KU*

c Describe how the diagram illustrates the antiparallel structure of DNA molecules. [1] *Applying KU*

d The diagram shows synthesis of the leading strand of DNA.
   Describe **one** difference between the replication of this strand and the other strand of the molecule. [1] *Demonstrating KU*

MARKS STUDENT MARGIN

2 The graph shows how temperature is changed during three stages in one cycle of a polymerase chain reaction (PCR).

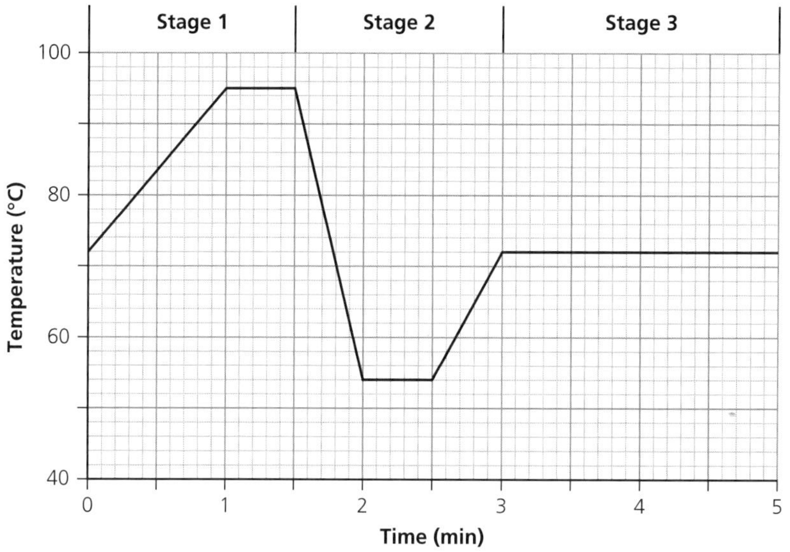

a State why the temperature is increased during stage 1. **1** Demonstrating KU

____

b Calculate the range of temperature that the PCR reaction tube experiences during one cycle of PCR. **1** Processing

Space for calculation

__________ °C

c Short sections of DNA called primers are involved in stage 2.

State the role of these primers during stage 2. **1** Demonstrating KU

____

____

d During stage 3, high temperatures would denature most enzymes but the polymerase used in this process remains active.

Explain why this is possible. **1** Demonstrating KU

____

____

e Explain the role of PCR in practical applications such as forensics. **1** Demonstrating KU

____

____

3 Give an account of tumour production by cancer cells. **4** Demonstrating KU

____

____

____

____

4 Hydrogen peroxide is a toxic chemical produced in human metabolism. Catalase is an enzyme that breaks down hydrogen peroxide.

hydrogen peroxide $\xrightarrow{\text{catalase}}$ water + oxygen

An experiment was carried out to investigate how the concentration of catalase affected the rate of hydrogen peroxide breakdown.

Filter paper discs soaked in different concentrations of catalase solution were added to beakers of hydrogen peroxide solution as shown in the diagram. The beakers were all kept at 20 °C throughout the experiment.

The discs sank to the bottom of the beakers before rising back up to the surface. The time taken for each disc to rise to the surface was used to measure the reaction rate. The faster the disc rises, the faster the reaction rate.
The results of the investigation are shown in the table.

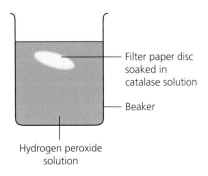

Catalase concentration (%)	Average time for ten discs to rise (s)
0.1	11.4
0.5	6.2
1.0	4.5
1.5	3.8
2.0	3.2
3.0	3.2

a **i** Using the information given, explain why the filter paper discs rose to the surface of the hydrogen peroxide solution.

ii Give the independent variable in this experiment.

iii Identify **one** variable that would have to be controlled to allow a valid conclusion to be made.

b Describe **one** feature of this experiment that helps to make the results more reliable.

c It was suggested that the filter paper itself reacted with the hydrogen peroxide.

Describe how the experiment could be controlled to allow this suggestion to be ruled out.

d On the grid below, plot a line graph to show the results of the investigation.

e Give **two** conclusions that can be made from the results of this experiment.

1 _____

_____

_____

2 _____

_____

_____

5 The diagram shows stages in the aerobic respiration of glucose.

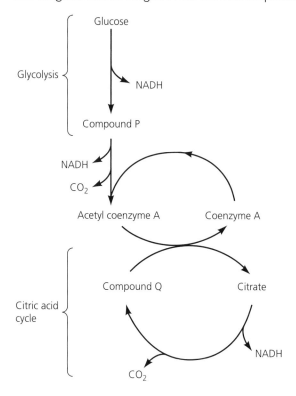

a Name compound P, which is a product of glycolysis.

b NADH is produced during these stages of respiration.
Name the stage of respiration into which the NADH passes and explain the importance of that stage.

Name

Explanation

c Name compound Q, which is regenerated in the citric acid cycle, and the location in cells where this metabolic pathway occurs.

Name

Location

d Describe the role of dehydrogenase enzymes in the citric acid cycle.

PRACTICE EXAM B

6 The diagrams represent sections through an ovary and part of a testis.

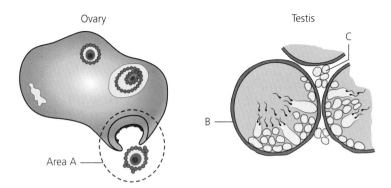

a Name the process that can be seen in area A and describe the role of a named hormone in its occurrence.

Name

_____

Description

_____

_____

b Name structure B and describe its role in reproduction.

Structure B

_____

Role

_____

_____

c Describe the role of the cells shown at C.

_____

_____

7 A study into diet and breast cancer was carried out involving 24 different countries in the world. For each country the average percentage fat in the diet of the population was plotted against the death rates for breast cancer and the results are shown in the graph.
A line of best fit was calculated and is also shown.

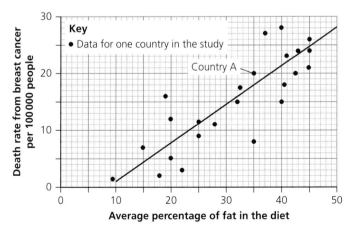

a Use values from the graph to describe the data for country A.

_____

_____

b It was concluded that risk of breast cancer is related to the percentage of fat in the diet.

Evaluate this statement by providing one piece of evidence from the graph that supports the statement and another which does not support it.

Supports

_____

_____

Does not support

_____

_____

c The study was extended to another country in which the average percentage of fat in the diet was 30%.

Use information from the graph to predict the expected death rate from breast cancer in this country.

_____ deaths per 100 000 people

8 The diagram shows a vertical section through a human brain.

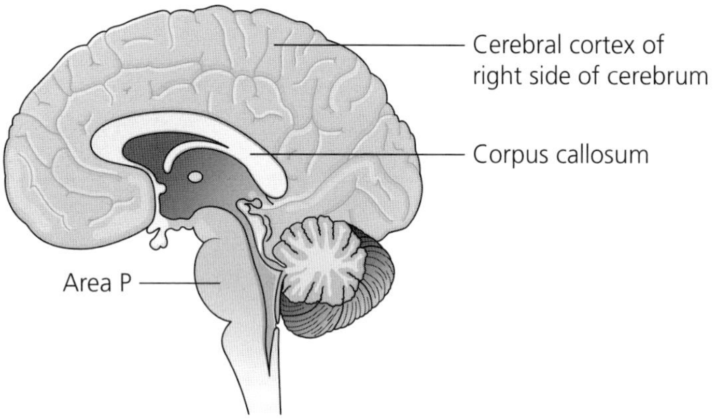

- Cerebral cortex of right side of cerebrum
- Corpus callosum
- Area P

a i Describe the function of the cerebral cortex of the **right** side of the cerebrum.

_____

_____

ii The cortex contains association areas.

Give **one** function with which association areas are involved.

_____

b Suggest how damage to the corpus callosum might affect the functioning of the brain.

_____

_____

c Give the function of area P.

_____

9 The diagram shows part of a neural pathway that controls the action of the hand. The arrows show the direction of the nervous impulses.

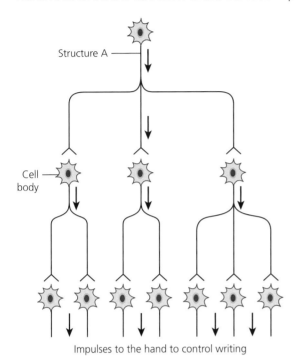

Impulses to the hand to control writing

a Name the type of neural pathway illustrated by the diagram.

b Describe how this pathway could provide fine motor control such as the use of the hand in writing.

c Name structure A shown in the diagram.

d Nervous impulses pass between neurons by chemical transmission at synapses.

    i Explain why neurotransmitter chemicals must be removed from the synaptic clefts.

    ii Describe the effect of an injection of an antagonistic drug on the neural pathway in the diagram.

PRACTICE EXAM B

10 The diagram represents the structure of an influenza virus.

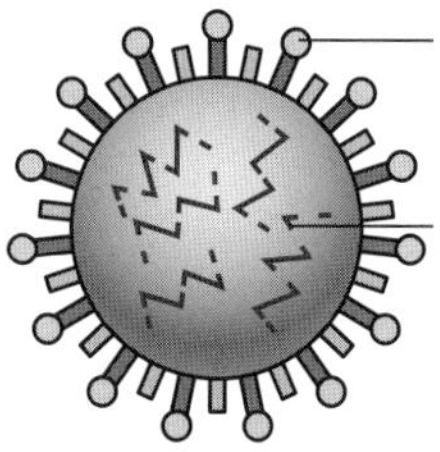

- Surface protein
- Nucleic acid

a In the preparation of a vaccine from this virus, the nucleic acid is destroyed but the surface protein molecules are left intact.

Name the surface proteins and explain why they must remain intact.

Name

_____

Explanation

_____

b Different vaccines are needed for different influenza strains.
Explain why the different vaccines are needed.

_____

c Some vaccines have aluminium hydroxide added to enhance their activity.
Give the term used for substances such as aluminium hydroxide in the production of vaccines.

_____

11 A healthy individual aged 25 years carried out a 6-month training programme.

Graph 1 shows the effect of standard exercise on the individual's heart rate before and after the training programme.

Graph 2 shows the relationship between the individual's stroke volume and heart rate before and after the training programme.

Graph 1

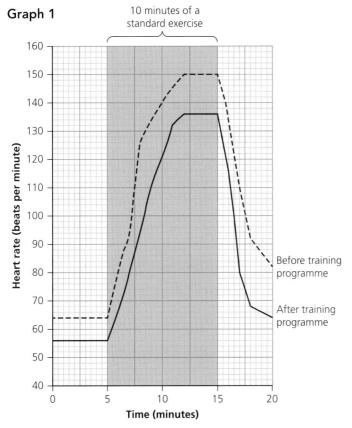

Graph 2

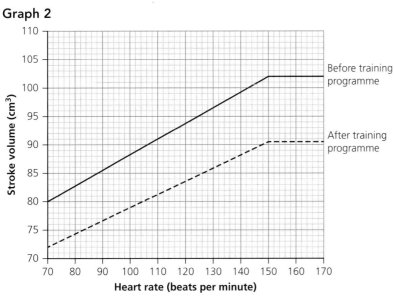

PRACTICE EXAM B

a Using data from **Graph 1**:

   i Express as a simple whole number ratio the individual's resting heart rate before and after the training programme.

   Space for calculation

   _____ : _____
   Before training   After training

   ii Calculate the difference in heart rate of the individual at the end of 10 minutes of a standard exercise before and after the training programme.

   Space for calculation

   _____ bpm

   iii Calculate the number of minutes into the standard exercise at which the training programme has made the most difference to the individual's heart rate.

   Space for calculation

   _____ minutes

b Use values from **Graph 2** to describe how stroke volume is affected when heart rate changes before the training programme.

   _____
   _____
   _____
   _____

c From **Graphs 1** and **2**:

   Calculate the cardiac output of the trained athlete 5 minutes after the start of the exercise.

   Space for calculation

   _____ cm³ per minute

MARKS	STUDENT MARGIN
1	Processing
1	Processing
1	Processing
2	Selecting
1	Processing

94

PRACTICE EXAM B

12 The diagram shows a synapse in a leg muscle of a sprinter.

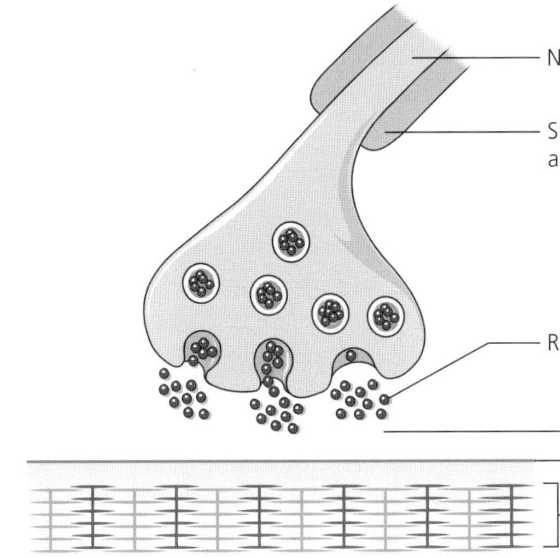

Diagram labels: Nerve fibre; Sheath of material around nerve fibre; Release of acetylcholine; Synaptic cleft; Post-synaptic membrane; Skeletal muscle fibre

a Name the material that would be found in the sheath around the nerve fibre and explain its role in the transmission of nerve impulses.

Material _____

Role _____

[2 marks — Applying KU]

b i Describe how acetylcholine affects the post-synaptic membrane.

_____

[1 mark — Demonstrating KU]

ii Explain how neurotransmitters are removed from the synaptic cleft following their action.

_____

[1 mark — Demonstrating KU]

iii Apart from acetylcholine, name **one** neurotransmitter substance.

_____

[1 mark — Demonstrating KU]

c Name the type of skeletal muscle fibre which would be expected to be most common in the leg muscles of sprinters.

_____

[1 mark — Demonstrating KU]

13 The diagram shows part of an electrocardiogram (ECG) trace obtained from a resting hospital patient.

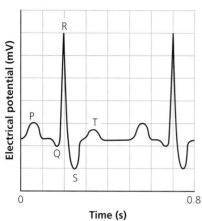

a i Using the information in the extract, calculate the resting heart rate of the patient.

Space for calculation

_____ beats per minute

   ii Describe the part of the cardiac cycle that is represented by the changes in voltage between points Q and R on the trace.

   _____

   _____

b Describe what is meant by the term diastole.

   _____

   _____

14 A study into the smoking habits of adults was carried out in a large Scottish town. The graph shows the percentages of adults in different categories of the town's population who smoked.

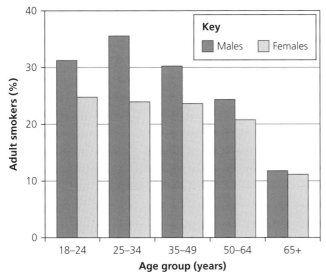

a Describe **two** differences between the trends shown in the graph for males compared with females.

1 _____

_____

2 _____

_____

b The nicotine in cigarette smoke causes an increase in dopamine levels in the blood and this can lead to addiction. In spite of this, the graph shows that the percentage of adults over 25 who smoke reduces with their age.

Give **two** suggestions which could explain the reduction in the percentage of smokers with age.

1 _____

_____

2 _____

_____

c Explain why repeated exposure to nicotine can lead to nicotine tolerance by desensitisation.

_____

_____

PRACTICE EXAM B

15 The diagram shows a T lymphocyte binding to a pathogen and part of the immune response which follows.

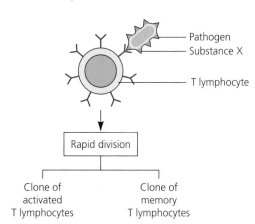

a Name substance X and describe the part substances such as this play in the immune response.

Substance X _____

Description _____

_____

[2 marks — Applying KU]

b i Describe what is meant by a clone, as shown in the diagram.

_____

[1 mark — Demonstrating KU]

ii Describe the action of the activated T lymphocytes.

_____

[1 mark — Demonstrating KU]

c Give **two** advantages to the individual of the existence of memory cells.

1 _____

_____

2 _____

_____

[2 marks — Demonstrating KU]

16 Three groups of volunteers took part in a double-blind clinical trial into the effect of drugs which inhibit the action of the parasympathetic and sympathetic nervous systems.

The members of the groups were given treatments as shown in the table. Data on arterial blood pressure and heart rate was collected during a period of time following treatment. The graph shows the data for the three groups.

Group	Treatment given
1	drug to inhibit the action of the parasympathetic nervous system
2	placebo
3	drug to inhibit the action of the sympathetic nervous system

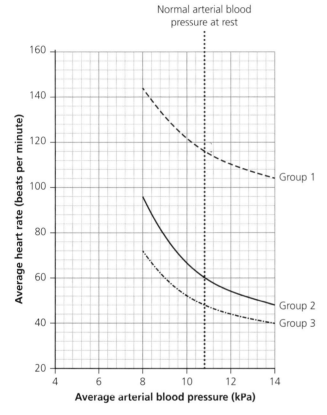

a   i Explain what is meant by a double-blind trial.

_____

_____

ii Group 2 received a placebo.
Describe what is meant by a placebo.

_____

_____

iii In this experiment, arterial blood pressure is measured in kPa.
Give one other unit which can be used to measure blood pressure.

_____

b Calculate the difference between the average heart rate of patients in Group 1 and those in Group 2 at normal arterial blood pressure at rest.
Space for calculation

_____ beats per minute

c From the results in the graph, give **one** conclusion which could be made about the control of heart rate by the sympathetic nervous system.

1 — Concluding

d Explain how blood pressure is involved in the creation of tissue fluid.

1 — Demonstrating KU

17 Answer **either question A or B** in the space below.
Labelled diagrams may be used where appropriate.
  A Discuss procedures that can be used to treat infertility.
  OR
  B Discuss the screening and testing procedures which may be carried out as part of antenatal care.

10 — Demonstrating KU

[End of Practice Exam B]

ANSWERS TO PRACTICE EXAMS

Practice Exam A
Paper 1

	Objective test			
Question	Answer	Mark	Demand	Commentary with hints and tips
1	D	1	C	One strand runs from 3' at the deoxyribose end to 5' at the phosphate end and the complementary strand runs in the opposite direction.
2	D	1	C	The answer here is about applying the complementary base pairing rules of **A**denine pairing with **T**hymine and **G**uanine pairing with **C**ytosine. 10% = C so 10% = G. This leaves 80%, so A = 40% and T = 40%. Ratio of C : T is therefore 10% : 40% = 1 : 4.
3	A	1	C	Diploid means two sets of chromosomes and haploid means one set. Meiosis reduces the diploid number to haploid whereas mitosis maintains the diploid number.
4	D	1	C	Typical translation diagram showing synthesis of polypeptide with tRNA transporting specific amino acids to the mRNA on the ribosome.
5	D	1	C	92 °C separates DNA strands 60 °C allows primers to attach 70 °C optimum temperature for DNA polymerase.
6	C	1	A	75% of maximum of 32 = 24 (reaction rate). Draw line across from 24 until it intersects graph then take line down to x-axis = 3.2.
7	C	1	C	Feedback inhibition is competitive so that the inhibitor can have a variable and reversible effect on the enzyme.
8	A	1	C	Remember that in the absence of an enzyme, the activation energy needed to start a metabolic reaction is greater.
9	C	1	A	The reaction will proceed to BQA, which will increase in concentration. Maleylacetoacetate is no longer produced so will decrease in concentration.
10	A	1	C	**ICSH** = **H**ormone that **S**timulates the **I**nterstitial **C**ells found between the seminiferous tubules.
11	B	1	C	Find the decrease in sperm cell count = 45 million Then divide the answer by the number of years of the study = 60 45/60 = 0.75 million (per cm^3)
12	D	1	C	The artery's endothelium is surrounded by a thick muscular wall, surrounded by an elastic outer wall.
13	A	1	C	Plasma proteins are too large to be forced out of the capillaries and so remain in the blood plasma.
14	A	1	A	75 + 45/3 75 + 15 = 90 mmHg
15	C	1	A	Remember that because HDLs transport cholesterol to the liver for elimination, a higher ratio of HDL : LDL is better for health because removing excess cholesterol decreases the risk of atherosclerosis.
16	B	1	A	Tricky! Read across from 40% of mothers to the pale grey line; then read down to the dark grey line for infant mortality; then read across to the infant mortality, which would be 5.2 babies per 100 000. 5.2 as a percentage of 100 000 is 5.2/100 000 × 100 – you may need to use your calculator to get 0.0052%.
17	A	1	C	The effects of oestrogen increase the chances of fertilisation.

ANSWERS TO PRACTICE EXAM A

Objective test

Question	Answer	Mark	Demand	Commentary with hints and tips
18	A	1	C	It is the build-up of phenylalanine which causes the effects of PKU so it follows that the enzyme which breaks it down does not function normally. PKU is caused by a substitution mutation.
19	B	1	C	Remember that the sympathetic system prepares the body for action whereas the parasympathetic system prepares for rest and allows energy to be diverted to activities such as digestion.
20	A	1	C	Agonists decrease receptor number and cause desensitisation leading to drug tolerance – try to learn this as a package.
21	C	1	C	Displacement is loss of information from short-term memory caused by further information entering before the existing information can be transferred.
22	C	1	A	Read across from 1 million deaths to the dotted line then up to the solid line for infections; from there read across to the scale at 23 million.
23	B	1	C	Adjuvants enhance the effect of vaccines.
24	C	1	A	Find the concentration after 10 days for first exposure = 5 units. Find the concentration after 10 days for second exposure = 13 units. Then find the percentage increase = 8/5 × 100 = 160%
25	B	1	C	The idea of herd immunity is that an unvaccinated individual is less likely to come into contact with an infected individual. Unvaccinated individuals may be unable to be vaccinated for a number of reasons.

Paper 2

Question		Expected answer	Marks	Demand	Commentary with hints and tips
1	a	Differentiation	1	C	Changes to cells that allow them to specialise for different functions. Specialised cells express the genes characteristic of that cell type.
	b	Tissue stem cells differentiate into a limited range of cell types/are multipotent = **1** Embryonic stem cells differentiate into (almost) any cell type/are pluripotent = **1**	2	CC	Knowing the terms pluripotent and multipotent makes this question easier to answer. Remember that you must give information for each of the two cell types here because the question asked for a description of the difference between them.
	c	Description: (cancer cells) divide excessively to produce a mass of abnormal cells = **1** Explanation: (cancer cells) fail to attach to each other **AND** spread through the body (to form secondary tumours) = **1**	2	CA	Just need to learn the stages of tumour and secondary tumour formation.

ANSWERS TO PRACTICE EXAM A

Question			Expected answer	Marks	Demand	Commentary with hints and tips
2	a		Transcription	1	C	The name given to the copying of DNA sequences to make a primary transcript.
	b		A triplet is a sequence of three nucleotides/bases **AND** each triplet encodes a specific amino acid	1	A	Triplets of bases on DNA and mRNA are called codons. Each triplet/codon codes for a specific amino acid.
	c		RNA polymerase	1	C	Make sure you mention which polymerase is involved – no marks will be awarded for polymerase alone.
	d		(Splice site mutations) can result in introns being left in the primary transcript **OR** exons being left out of the primary transcript	1	C	Mutation at a point where coding and non-coding regions meet in a section of DNA. A single gene mutation at a splice site could result in an intron being left in the mature mRNA and so contributing to protein structure.
3	a		Missense = **1** Protein/polypeptide formed is shorter/contains fewer amino acids than it should = **1**	2	CC	Remember: • **m**issense = **m**istake in a single amino acid • **n**onsense = **n**o further translation, i.e. a new stop codon.
	b	i	A part of a chromosome is removed **AND** becomes attached to another non-homologous chromosome	1	A	Make the names and definitions of the chromosome mutations into flashcards to help you learn and remember them.
		ii	Deletion **OR** Duplication **OR** Inversion	1	C	Remember that duplication occurs when a piece of chromosome becomes attached to its homologous partner.
4	a		Q = **1** P = **1**	2	CC	'Before and after' shows substrate molecule P broken into products S and T. Inhibitor Q acts as a non-competitive inhibitor, attaches to a site on the enzyme other than the active site, changing the shape of the active site.
	b		Non-competitive = **1** Inhibitor molecule binds to the enzyme molecule **AND** changes the shape of the active site = **1**	2	CA	Enzyme inhibition by a substance that binds away from the active site but permanently alters the active site of the enzyme.
5	a	i	3.0 mM per litre	1	A	Take care with these double y-axis graphs. Use a ruler to intersect the correct plot and then make sure you read the value from the correct axis.

ANSWERS TO PRACTICE EXAM A

Question			Expected answer	Marks	Demand	Commentary with hints and tips
		ii	0.3 mM per litre	1	A	These 'average increase' questions are often poorly done by candidates. First calculate the increase and then divide by the number of seconds. 7 − 1 = 6 then 6/20 = 0.3 mM per litre
		iii	1.5 mM per litre	1	A	Use a ruler to draw a straight line across from 5 mM per litre on the right y-axis until it intersects the lactate plot. Next, draw a line down until it intersects the creatine phosphate plot. Then draw a line across to the left y-axis to obtain the value of the creatine phosphate.
	b		Creatine phosphate: When activity becomes strenuous, more ATP is required and creatine phosphate is broken down to release phosphate and energy for its synthesis = **1** Lactate: When activity becomes strenuous, oxygen becomes deficient in muscle cells and some pyruvate is converted to lactate = **1**	2	AC	Creatine phosphate (CP) breaks down to release energy and phosphate, which are used to convert ADP to ATP at a fast rate, and therefore help support strenuous exercise in muscle cells. In oxygen-deficient conditions, pyruvate from glycolysis is converted to lactate.
	c	i	Untrained individual: Middle-distance runner has more slow-twitch **AND** less fast-twitch than the untrained individual **OR** Converse = **1** Elite sprinter: Middle-distance runner has less slow-twitch **AND** more fast-twitch than the elite sprinter **OR** Converse = **1**	2	CA	Remember: **SS** = **S**low-twitch for '**S**tamina'. Straightforward 'describe' question, selecting information and values if necessary from the data provided.
		ii	High % of fast-twitch is suitable for the rapid bursts of strenuous activity needed to lift a heavy weight	1	C	Remember: Fast-twitch for 'Fast and Furious', e.g. sprint and power. They contract more quickly and powerfully. **FF** = **F**ast-twitch **F**atigue quickly.
6	a		Glycolysis	1	C	It is essential to remember the names of the three stages of aerobic respiration.
	b		NADH **OR** Hydrogen ions **AND** electrons	1	C	Remember that you must mention the electrons as well as the hydrogen ions.

ANSWERS TO PRACTICE EXAM A

Question			Expected answer	Marks	Demand	Commentary with hints and tips
	c		Substance S: oxygen = **1** Role: final acceptor of hydrogen (ions) **AND** electrons = **1**	2	CA	Important to mention that oxygen is the final acceptor and that it accepts electrons as well as hydrogen ions.
7	a		Luteinising hormone/LH	1	C	Pituitary releases **F**SH **F**irst and **L**H **L**ast.
	b	i	Proliferation/thickening/further development/vascularisation of the endometrium/uterus	1	C	Progesterone promotes the further development and vascularisation of the endometrium, preparing it for implantation of a blastocyst if fertilisation occurs.
		ii	Breakdown of uterus wall leading to menstruation	1	C	A drop in level triggers menstruation when fertilisation has not occurred.
	c		Pituitary = **1** Stimulates development/maturation of a follicle **OR** stimulates production of oestrogen (by the follicle) = **1**	2	AC	FSH initiates gamete release through the development of follicles in the ovaries and the production of oestrogen from the follicles.
8	a		She has inherited h from her father (who is X^hY) **AND** from her mother who is a carrier/is X^HX^h/is heterozygous	1	A	Difficult. A mark that requires candidates to describe both the male and the female. The answer should refer to the single X chromosome in males and the double XX chromosomes in females.
	b		X^HX^h	1	A	S must be heterozygous. She has received the X^h from her father and since she does not show the condition she must have received an X^H from her mother.
	c	i	Meaning: Method to obtain embryonic cells from amniotic fluid before birth (which can be cultured) = **1** Explanation: Sex chromosomes in karyotype resulting from culture of amniotic cells can be examined before birth = **1**	2	CA	Make flashcards for all of the antenatal and postnatal screening procedures and terms and get someone to test you. A karyotype is an image of an individual's chromosomes, arranged in homologous pairs. A karyotype is used to identify anomalies in the numbers or structure of chromosomes but can also reveal the sex of the fetus.
		ii	100% (X^hY)	1	C	A male child will receive their single X chromosome from their mother. Since the mother has the condition she must be X^hX^h and so the male child must receive an X^h and will have the condition.

ANSWERS TO PRACTICE EXAM A

Question			Expected answer	Marks	Demand	Commentary with hints and tips
9	a	i	Some patients had mean insulin sensitivity within the range for type 2 diabetic patients following GBS	1	A	To answer this, you need to apply the ranges provided to obtain the upper and lower values. Then compare the non-diabetic range with the type 2 diabetes range.
		ii	Some patients had greater sensitivity to insulin following GBS = **1** Some patients had lower sensitivity to insulin following GBS = **1**	2	CA	Again, you need to apply the ranges provided to obtain the upper and lower values. Then compare the type 2 diabetes range with the non-diabetic range to see the differences.
	b		In non-diabetic individuals, insulin makes the liver cells more permeable to glucose = **1** Activates the conversion of glucose to glycogen = **1**	2	CA	Insulin makes the liver cells more permeable to glucose and activates the conversion of glucose to glycogen, decreasing the blood glucose concentration. Both permeability and storage must be mentioned.
10	a		Physical injury **OR** infection **OR** accumulation of low-density lipoproteins	1	C	If an atheroma ruptures, the damage to the endothelium causes the release of clotting factors.
	b		Fibrinogen	1	C	Try to memorise the clotting process by writing the events as a flowchart.
	c		Thrombus detaches from the site of thrombosis = **1** Becomes trapped and blocks a coronary artery = **1**	2	CA	Thrombosis is the formation of a blood clot (thrombus) in a vessel. If a thrombus breaks loose, it forms an embolus that travels through the bloodstream until it blocks a blood vessel. The blood vessels which supply heart tissues are the coronary arteries.
11	a		Medulla	1	C	Autonomic nervous system is located in the medulla.
	b		Sympathetic nerve increases heart rate **AND** parasympathetic nerve decreases heart rate	1	A	Sympathetic nerve is the 'accelerator' – speeding it up. Parasympathetic nerve is the 'brake' – slowing it down.
	c		AVN/atrioventricular node	1	C	AVN is between the atria and the ventricles.
	d		AVN receives impulses from the SAN = **1** Passes impulses down the fibres Y, which stimulate contraction of the ventricle walls during ventricular systole = **1**	2	CA	This results in the simultaneous contraction of the ventricles (ventricular systole).

ANSWERS TO PRACTICE EXAM A

Question			Expected answer	Marks	Demand	Commentary with hints and tips
12	a		X: aorta = **1** Y: SL/pulmonary valve = **1**	2	CC	Straightforward recall of heart structure.
	b	i	Volume of blood pumped out of chamber R/chamber Q in one beat	1	C	Cardiac output (CO) = stroke volume (SV) × heart rate (BPM)
		ii	Stroke volume × number of beats per minute	1	C	
13			1 Carried out before birth 2 Dating scan shows pregnancy stage/due date 3 Anomaly scans to detect physical abnormalities 4 Blood and urine tests to monitor concentrations of marker chemicals 5 Amniocentesis/chorionic villus sampling (CVS) 6 Samples cells to be cultured to produce a karyotype 7 CVS can be done earlier but carries greater risk of miscarriage **(Any four – 1 mark each)**	4	CCCC	This answer relies on remembering a set of related and coherent facts about antenatal screening.
14	a		X: axon = **1** Function: Insulates fibres **AND** speeds up transmission of impulses = **1**	2	CC	**AA** = **A** for **A**xon, **A** for impulses **A**way from the cell body. Myelination continues from birth to adolescence.
	b		Gene for myelin is switched on/able to be transcribed	1	A	During differentiation, genes which express proteins characteristic of the differentiated cell are switched on. Remember that answers might need you to use information from more than one Key Area.
	c		Vesicle fuses with membrane and releases transmitter into synaptic cleft = **1** Transmitter crosses synapse and binds to receptors in post-synaptic membrane (to pass on impulse) = **1**	2	CA	Neurotransmitters bind with receptors. It is the type of receptor involved that determines whether a signal is excitatory or inhibitory.
15	a		(That increased) caffeine level in blood improves/reduces/has no effect on learning (of motor tasks)	1	C	When an observation is made, the suggested scientific explanation for it is called a hypothesis – a possible explanation made as a starting point for further investigation.

ANSWERS TO PRACTICE EXAM A

Question		Expected answer	Marks	Demand	Commentary with hints and tips
	b	Average number of errors per group	1	C	The **d**ependent variable refers to the **d**ata (results) produced.
	c	Diet up to the time of the experiment **OR** Time between taking caffeine and trial **OR** External stimuli other than the maze task **OR** Method of administering the caffeine drink **OR** Concentration of the caffeine drink **OR** Gender balance/health of the participants	1	C	If other variables are not controlled, differences in results could be due to them and not to the independent variable under investigation.
	d	Repeat the experiment exactly but with another group of ten 25-year-old volunteers and no additional caffeine	1	C	The control should be identical to the original experiment apart from the independent variable being investigated.
	e	Axes scaled **AND** labelled correctly with units = **1** Points plotted accurately, connected with straight lines **AND** correct key **OR** labelled lines = **1**	2	CC	Marks are given for providing scales, labelling the axes correctly and plotting the data points. Line graphs require points to be joined with straight lines using a ruler. The graph labels should be identical to the table headings and must include units. Choose scales that use at least half of the graph grid provided, otherwise a mark will be deducted. The values of the divisions on the scales you choose should allow you to plot all points accurately. Make sure that your scales include zero if appropriate and extend beyond the highest data points. The scales must rise in regular steps.
	f	Caffeine increases the rate of learning	1	C	The more caffeine taken in the experiment, the fewer errors made and the faster the errors cease. It is essential to relate the conclusion to the aim of the experiment, which is mentioned in the stem of the question, and not simply to relate caffeine to errors.
	g	Repeat the experiment with more individuals	1	C	To improve the reliability of an experiment and the results obtained, the experiment should be repeated. Remember **ROAR**: **r**epeat and **o**btain an **a**verage, which increases **r**eliability.

ANSWERS TO PRACTICE EXAM A

Question			Expected answer	Marks	Demand	Commentary with hints and tips
16	a		Cerebrum	1	C	Not to be confused with the cerebellum!
	b		Different regions marked are active in carrying out different specific functions	1	A	This suggests that different areas are dedicated to different types of activity.
	c	i	The motor area is controlling the actions involved with folding the paper = **1** The sensory area is collecting audio information about the description being read out loud = **1**	2	CA	Linking up the parts of the task with the areas requires you to know that motor areas control muscles and sensory areas receive input from sense organs such as the eyes.
		ii	Association	1	C	You only have to be aware that there are motor, sensory and association areas in the cerebrum.
17	a		Divide into groups in a randomised way = **1** to reduce bias because of age or gender = **1**	2	CA	Random allocation to groups can go some way to ensuring that the groups can be compared validly.
	b		The two groups must be of a suitable size	1	C	Suitable sizes are as large as is practicably possible. The larger the sample size the more reliable the data collected will be.
18	a		91–92 increased from 2800 to 3400 reported cases 92–93 decreased from 3400 to 2600 reported cases 93–96 increased from 2600 to 3600 reported cases **(Any two – 1 mark each)**	2	CA	Always remember to quote key values and units.
	b		18%	1	A	First, calculate the increase or decrease. Then, express this value as a percentage change by dividing by the original starting value then multiplying by 100.
	c		Good hygiene/hand washing following use of toilets **OR** Appropriate waste disposal systems/quality water supply **OR** Good hygiene/hand washing before preparation of food **OR** Appropriate storage/handling of food **OR** Treating affected individuals (with antibiotics)	1	C	A 'suggest' question, with lots of possible answers.
	d		4000	1	C	Make the assumption that the latest trend of increase between 1994 and 1996 will be continued into 1997.

Question		Expected answer	Marks	Demand	Commentary with hints and tips
19	A	1 Epithelial cells/skin form a physical barrier 2 Epithelial cells produce secretions against infection 3 Mast cells produce histamine 4 Histamine produces inflammation/vasodilation/capillary permeability/increased blood flow 5 Phagocytes/natural killer/NK cells release cytokines which stimulate the specific immune response 6 Cytokines lead to accumulation of phagocytes at infection sites 7 Cytokines lead to delivery of antimicrobial proteins to infection sites 8 Cytokines lead to delivery of clotting elements to infection sites 9 Phagocytes recognise surface antigens on pathogens 10 In phagocytosis, pathogens are engulfed/rendered harmless 11 Natural killer/NK cells induce virally infected cells to undergo apoptosis/produce self-destructive enzymes (Any 8)	8	CCCCCCAA	In full extended response questions, it is vital to use the language that is in the course specification – this is the key to scoring marks. Having a few extended response answers for each Area of study well-rehearsed is a good exam revision technique.

ANSWERS TO PRACTICE EXAM A

Question		Expected answer	Marks	Demand	Commentary with hints and tips
	B	1 Lymphocytes respond specifically to antigens on foreign cells/pathogens/toxins released by pathogens 2 T lymphocytes have specific surface proteins that allow them to distinguish between body cells and cells with foreign molecules on their surface 3 T lymphocytes induce apoptosis 4 T lymphocytes secrete cytokines which activate B lymphocytes and phagocytes 5 B lymphocytes produce specific antibodies 6 Antibodies travel in blood and lymph to infected areas 7 Antibodies recognise antigens 8 Antigen–antibody complexes inactivate pathogens or render them more susceptible to phagocytes 9 Antigen–antibody complexes stimulate cell lysis 10 Some T and B lymphocytes survive as memory cells 11 Secondary exposure to antigens results in more rapid and greater immune response (Any 8)	8	CCCCCCAA	In full extended response questions, it is vital to use the language that is in the course specification – this is the key to scoring marks. Having a few extended response answers for each Area of study well-rehearsed is a good exam revision technique.

Practice Exam B
Paper 1

	Objective test			
Question	Answer	Mark	Demand	Commentary with hints and tips
1	D	1	C	Tissue (adult) stem cells in bone marrow differentiate into red blood cells, platelets and the various forms of phagocytes and lymphocytes.
2	B	1	A	Max = 8 so 50% = 4. Use your ruler to draw a line along from the 4 on the left axis until it intersects the stem cell line, then draw a line down until it intersects the glutaminase line. Finally, draw a line over to the right axis to obtain the answer.
3	B	1	C	Remember that tRNA molecules are single stranded but have hydrogen bonds between bases to hold them into a definite shape.
4	D	1	C	Primary transcript is transcribed from DNA and introns are the non-coding regions which are later spliced out to give mature mRNA.
5	D	1	C	The shape of the inhibitor does not allow binding to the active site, which is occupied by the substrate, so it must be a non-competitive inhibitor and so not affected by changing the substrate concentration.
6	C	1	C	Missense and nonsense are also substitutions. Deletion changes the base order in all codons after the deletion and the effects frameshift from there.
7	B	1	C	Straightforward – remember that drugs are pharmaceutical products.
8	B	1	C	**SSMM** = **S**low-twitch for **S**tamina events, **M**itochondria and **M**yoglobin. **FF** = **F**ast-twitch for **F**ast events. Anaerobic and only involves glycolysis.
9	C	1	A	These 'average increase' questions are often poorly done by candidates. First calculate the increase then divide by the time period, in this case 20 seconds. 8.0 – 0.6 = 7.4 then 7.4/20 = 0.37 mM per litre
10	C	1	C	This involves rearranging the formula: SV = CO/HR
11	D	1	C	Simple question – remember that the coronary artery branches off the aorta and spreads over the heart's outer surface, supplying the cardiac muscle with oxygen and nutrients.
12	A	1	A	High-density lipoproteins (HDL) transport excess cholesterol from the body cells to the liver for elimination. Low-density lipoproteins (LDL) transport cholesterol to body cells.
13	C	1	A	First take care that you are familiar with the scales on each axis. A concentration of 6 units of insulin/cm^3 is present when the concentration of glucose is 117 mg for every 100 cm^3. So in 4.8 litres of blood there are: 4800/100 × 117 = 5616 mg of glucose.
14	A	1	C	A person with type 1 diabetes is unable to produce insulin and is treated with regular injections of insulin. Individuals with type 2 diabetes produce insulin but their cells are less sensitive to it. This is linked to a decrease in the number of insulin receptors on the liver cells.

ANSWERS TO PRACTICE EXAM B

Objective test

Question	Answer	Mark	Demand	Commentary with hints and tips
15	A	1	C	Evidence indicating recessive because neither parent shows the condition and autosomal because if it was sex-linked the male parent would have to show the condition for any daughters to have the condition.
16	A	1	C	Key is to note that the techniques must be about obtaining cells.
17	B	1	C	Try making up flashcards with the name of the neurotransmitter on one side and the functions on the reverse.
18	D	1	C	Just requires careful reading of each statement to see if it applies to the data provided.
19	D	1	A	Number required/Total × 100 So, add up number of deaths not caused by infection (12) then divide by total number of deaths (20) then multiply by 100. 12/20 × 100 = 60%
20	B	1	C	Autoimmunity occurs when the body produces antibodies against its own cells.
21	A	1	C	Sympathetic is all about preparing the body for action and parasympathetic is about preparing for rest.
22	C	1	C	If information enters the short-term memory but is not transferred to long-term memory, then it must have been displaced.
23	B	1	C	This is the definition of cytokines given in the course specification.
24	B	1	C	Common harmless antigens that are involved with allergy include items such as pollen and peanuts.
25	D	1	C	Need to consider the error bars and note if they overlap or not. If they do the data is not significantly different.

Paper 2

Question		Expected answer	Mark	Demand	Commentary with hints and tips
1	a	Phosphate	1	C	Remember that the DNA backbones are made of sugars and phosphates linked – the sugars are at the 3' ends on the strands.
	b	Thymine	1	C	The answer here is about applying the complementary base pairing rules of **A**denine pairing with **T**hymine and **G**uanine pairing with **C**ytosine.
	c	The 3' (deoxyribose) end of one strand is bonded/joined to the 5' (phosphate) end of its complementary strand	1	A	Parallel strands in DNA but running in opposite directions. Must mention the 3' and 5' ends shown in the diagram.

ANSWERS TO PRACTICE EXAM B

Question			Expected answer	Mark	Demand	Commentary with hints and tips
		d	The lead strand is replicated continuously **AND** the lagging strand/other strand is replicated/built up in sections/fragments	1	A	A primer joins the end of the 3'–5' leading template strand and DNA polymerase adds free DNA nucleotides to synthesise a complementary strand continuously. On the lagging strand, primers are added one by one into the replication fork as it widens and DNA nucleotides are added to form fragments. These fragments are then joined by DNA ligase to form a complete complementary strand.
2	a		To separate the (DNA) strands/break the hydrogen bonds between strands/denature the DNA	1	C	Remember: • 92–98 °C separates DNA strands • 55–65 °C allows primers to attach • 70 °C optimum for heat-tolerant DNA polymerase.
	b		41 °C	1	C	Find the upper and lower values then subtract to find the range: 95 − 54 = 41
	c		They bind/anneal/join to (the ends of the) target/complementary sequences (of DNA being copied)	1	C	Primers are short pieces of single-stranded DNA that are complementary to the target sequence. The DNA polymerase begins synthesising new DNA from the end of the primer.
	d		Enzymes used in this procedure are heat tolerant/from hot spring bacteria	1	C	*T. aquaticus* is a bacterium that lives in hot springs and hydrothermal vents, and so its enzyme **Taq polymerase** can withstand the protein-denaturing high temperatures required during PCR.
	e		Amplifies/makes many copies of DNA	1	C	PCR is now commonly used as part of a wide variety of applications, including genotyping, cloning, mutation detection, sequencing, forensics and paternity testing. Remember though, PCR only amplifies the DNA to give a big enough sample to work with.

ANSWERS TO PRACTICE EXAM B

Question			Expected answer	Mark	Demand	Commentary with hints and tips
3			1 Cancer cells have uncontrolled cell division/divide excessively 2 Do not respond to regulatory signals 3 Produce a mass of abnormal cells/which is a tumour 4 May fail to attach to each other 5 (If they fail to attach to each other) they can spread through the body 6 This can lead to secondary tumours **(Any four – 1 mark each)**	4	CCCA	This answer relies on remembering a set of related and coherent facts about tumour formation. You could present this information as a flowchart.
4	a	i	Oxygen is produced (becoming trapped in the filter paper, causing it to float)	1	A	Need to make the link between the information provided regarding oxygen gas/bubble release and how this could cause the paper discs to rise.
		ii	Catalase concentration	1	C	**In**dependent variables are being **In**vestigated. **D**ependent variables give the **D**ata that forms the results.
		iii	Volume hydrogen peroxide Concentration hydrogen peroxide Volume catalase pH Diameter/mass/type of filter paper Others **(Any one)**	1	C	Always look for volumes and concentrations; note that temperature is already given but pH is a possible answer.
	b		Ten discs were used at each concentration **OR** The experiment was repeated at each concentration **OR** Average time was taken for each concentration	1	C	**ROAR** = **R**epeat, **O**btain an **A**verage, **R**eliable. In this instance, we see that a table with average time has been included and so this means that the experiment was replicated to improve reliability.

115

ANSWERS TO PRACTICE EXAM B

Question		Expected answer	Mark	Demand	Commentary with hints and tips
	c	Use discs soaked in water (added to hydrogen peroxide) **OR** Use discs containing no catalase (added to hydrogen peroxide)	1	A	The control should be identical to the original experiment apart from the one factor being investigated. If you are asked to describe a suitable control, make sure you describe it in full. A control experiment allows a comparison to be made and allows you to relate the dependent variable to the independent one.
	d	Correct scales, labels and units on axes (average time (s) is acceptable) = **1** Points correctly plotted and line drawn = **1**	2	CA	Marks are given for providing scales, labelling the axes correctly and plotting the data points. Line graphs require points to be joined with straight lines using a ruler. The graph labels should be identical to the table headings, including units. Choose scales that use at least half of the graph grid provided, otherwise a mark will be deducted. The values of the divisions on the scales you choose should allow you to plot all points accurately. Make sure your scales include zero if appropriate and extend beyond the highest data points. The scales must rise in regular steps.
	e	**1** As (catalase) concentration increases, reaction rate/rate of hydrogen peroxide breakdown increases = **1** **2** At higher concentrations/above 1% the reaction rate levels off = **1**	2	CA	When concluding, you must refer to the experimental aim, which is likely to be stated in the stem of the question. This conclusion requires you to describe both clear trends shown.
5	a	Pyruvate	1	C	During glycolysis, glucose is broken down to pyruvate in the absence of oxygen.
	b	Electron transport chain = **1** Use energy from the flowing electrons to synthesise ATP = **1**	2	CA	The energy is used to pump hydrogen ions through the inner membranes of the mitochondria, then they flow back through ATP synthase making ATP at the same time.
	c	Oxaloacetate = **1** Matrix of mitochondria = **1**	2	CC	Oxaloacetate is regenerated from citrate by enzyme-mediated reactions in the citric acid cycle.
	d	Removes hydrogen ions **AND** electrons from/oxidises substrate	1	A	Dehydrogenases remove hydrogen ions and electrons from intermediates in the citric acid cycle. Remember to use the term hydrogen **ions**.

ANSWERS TO PRACTICE EXAM B

Question			Expected answer	Mark	Demand	Commentary with hints and tips
6	a		Ovulation = **1** High level of LH triggers the release of a (mature) ovum (from an ovary/into an oviduct) = **1**	2	CA	This is the event which marks the end of the follicular phase and the start of the luteal phase of the menstrual cycle.
	b		Seminiferous tubule = **1** Sperm production = **1**	2	CC	Testes are made up of coiled tubules, which show up as circular shapes in the diagram.
	c		Produce/release testosterone	1	C	Remember this by the 'T' sound – in**T**ers**T**i**T**ial cells produce **T**es**T**os**T**erone.
7	a		Has a diet with 35% fat on average = **1** Has a death rate from breast cancer of 20 per 100 000 = **1**	2	CA	This simply requires you to take the information/values straight from the x- and y-axes.
	b		As the average percentage of fat in the diet increases, the death rate from breast cancer increases = **1** Similar death rates at different average % fat in diet = **1**	2	CA	First statement supporting the conclusion is perhaps more obvious. Noticing the different death rates despite the same percentage of fat in the diet is more challenging.
	c		14.5	1	C	The most obvious prediction is best obtained from the line of best fit so read up from 30% to the best fit line then across to the y-axis scale.
8	a	i	Receives/processes information from the left side of the body **AND** sends information to the muscles/effectors on the left side of the body	1	A	Tricky. Make sure you mention both incoming and outgoing information.
		ii	Language processing **OR** personality **OR** imagination **OR** intelligence **(Any one)**	1	C	These are the association areas which are explicitly mentioned in the course specification so use them to be safe.
	b		Damage to the corpus callosum interrupts the flow of information from one cerebral hemisphere to the other	1	C	It is important to realise that the two sides of the brain work together and share information.
	c		Contains centres for/controls breathing and heart rate	1	C	Area P is the medulla.

ANSWERS TO PRACTICE EXAM B

Question			Expected answer	Mark	Demand	Commentary with hints and tips
9	a		Diverging/divergent	1	C	In diverging neural pathways, impulses from one neuron are passed to several others.
	b		Impulses go to many/a number of effectors/muscles/fingers	1	C	Allows several muscles to be activated/stimulated/contracted at once and brings about the coordinated fine motor movement of the hand needed for writing.
	c		Axon	1	C	**AA** = **A**xons conduct impulses **A**way from cell bodies.
	d	i	To maintain sensitivity/prevent continuous stimulation/allow system to respond to new signals	1	A	If the neurotransmitter was left in the synaptic cleft there would be continuous stimulation of the post-synaptic membrane and the system would not be able to respond to new signals, making precise control impossible.
		ii	Inhibits transmission of nerve impulse **OR** Binds to receptors in the synapses	1	C	Antagonistic drugs block the action of natural neurotransmitters and so prevent nerve impulses passing synapses.
10	a		Antigens = **1** To allow recognition by the immune system/lymphocytes **OR** So antibodies/memory cells can be produced = **1**	2	CA	Surface protein is an antigen – a molecule that can produce an immune response by the body.
	b		Different strains of flu/the viruses have different antigens/surface proteins/antigenic markers/show antigenic variation	1	C	Mutations can result in different surface proteins.
	c		Adjuvants	1	C	**Ad**juvants **Add** to the immune response that the vaccines trigger.
11	a	i	8 : 7	1	C	First you need to obtain the values for the ratio from the data provided in the graph. Take care that you present the ratio values in the order in which they are stated in the question. Then simplify them, by dividing the larger number by the smaller one and then dividing the smaller one by itself. If this does not give a whole number ratio, you need to find another number that will divide into both values; divide each side of 64 : 56 by 8 to simplify to 8 : 7.
		ii	14	1	C	From Graph 1: 150 – 136 = 14
		iii	3	1	C	Needs careful measurement.

ANSWERS TO PRACTICE EXAM B

Question			Expected answer	Mark	Demand	Commentary with hints and tips
	b		Between 70 and 150 bpm stroke volume increased from 80 cm³ to 102 cm³ = **1** From 150 to 170 bpm stroke volume remains steady/constant at 102 cm³ = **1**	2	CA	When you are asked to describe a trend, it is essential that you quote the values of the appropriate points and use the exact labels given on the axes in your answer. You must use the correct units in your description.
	c		10 020 cm³	1	A	Cardiac output = heart rate × stroke volume. From Graph 1: heart rate of individual 5 minutes after the start of the exercise period after training = 120. From Graph 2: a heart rate of 120 gives a stroke volume of 83.5 So CO = 120 × 83.5 = 10 020
12	a		Myelin = **1** Insulates the nerve fibre and so speeds up the nerve impulse = **1**	2	CA	It's better to mention insulation as the reason for the increase in the nervous impulse speed.
	b	i	Binds to specific receptors to allow the transmission of the nerve impulse	1	A	Specificity is an important part of the answer here.
		ii	Enzyme action **OR** reabsorption	1	C	You don't have to know which transmitter is removed by which method.
		iii	Noradrenaline **OR** dopamine **OR** endorphin	1	C	Others also possible.
	c		Fast-twitch fibres	1	C	**F**ast **f**ibres for **f**ast activities.
13	a	i	120	1	C	Note that one beat takes five boxes to complete; each box is 0.1 of a second so 5 × 0.1 = 0.5 s for one beat. Therefore 60/0.5 beats in 1 minute = 120 beats per minute.
		ii	Ventricular systole = **1** Blood forced out of the heart from the ventricles = **1**	2	CA	This is characterised by a large change in potential across the ventricles.
	b		Relaxation of heart muscle (between beats)	1	C	Systole is contraction and diastole is relaxation.
14	a		A higher percentage of males smoke compared with females = **1** The percentage of women who smoke decreases throughout the age period shown **AND** for males there is an increase in the percentage of smokers between 18–24 and 25–34 before a decrease is seen = **1**	2	CA	Make sure you compare the data for males and females in your answers.

119

ANSWERS TO PRACTICE EXAM B

Question			Expected answer	Mark	Demand	Commentary with hints and tips
	b		**1** more smokers die = **1** **2** more individuals stop smoking as age increases = **1**	2	CA	Just think of the ways in which an individual might no longer be smoking.
	c		Nicotine is an agonist which stimulates specific receptors = **1** causing the nervous system to decrease both the number and sensitivity of these receptors (leading to drug tolerance) = **1**	2	CA	Just have to learn the difference between agonist and antagonist drugs.
15	a		Antigen = **1** Triggers an immune response **OR** specifically binds to an antibody = **1**	2	CA	Remember to use the word specific in this type of answer.
	b	i	A group of identical lymphocytes produced by mitosis of an activated lymphocyte	1	A	Remember that mitosis produces genetically identical daughter cells.
		ii	Bind to antigens on infected cells **AND** release proteins which cause destruction of cell/trigger destruction of cell by enzymes	1	C	Remember – binding to antigens by membrane receptors on lymphocytes is specific, meaning only infected cells will be destroyed.
	c		Response/production of antibodies to a reinfection will be; faster **OR** greater **OR** more long-lasting **(Any two – 1 mark each)**	2	CA	It is the memory cells which allow the immune response to prevent an infection from developing into disease.
16	a	i	Method in which neither trial participant nor experimenter knows the treatment given	1	C	The purpose of this kind of study is to eliminate the power of suggestion or bias.
		ii	(Participants are given) a treatment which does not contain the drug (under trial)	1	C	Researchers use placebos during studies to help them understand what effect a new drug or some other treatment might have on a particular condition. They then compare the effects of the drug or treatment and the placebo on the people in the study.
		iii	mm Hg	1	C	The unit mm Hg is millimetres of mercury – the units used to measure blood pressure.
	b		56 (± 1)	1	C	Read up from the normal arterial pressure at 10.8 kPa and find where the line for Group 2 cuts = 60 bpm and where the line for Group 1 cuts = 116 bpm; then subtract the values.

ANSWERS TO PRACTICE EXAM B

Question		Expected answer	Mark	Demand	Commentary with hints and tips
	c	Sympathetic keeps heart rate up/higher/increases heart rate (compared with control)	1	C	Group 1 results show what happens in the absence of the parasympathetic system and so shows the effects of the sympathetic system clearly.
	d	Pressure filtration/high pressure forces plasma out of capillaries and into tissue fluid	1	A	Capillaries are leaky so pressure within them causes squeezing out of fluid.
17	A	1. Ovulation can be stimulated by drugs 2. These prevent the negative feedback of oestrogen on FSH production 3. Other drugs/hormones (not FSH/LH) can be given which mimic the action of FSH/LH 4. They cause super ovulation/the production of a number of ova/eggs 5. *In vitro* fertilisation/IVF programmes 6. Ova are removed (surgically) from the ovaries 7. Ova are mixed with sperm/fertilisation occurs outside the body 8. Fertilised eggs divide to form a ball of cells/at least eight cells form/form a blastocyst 9. Transferred into the uterus (for implantation) 10. Artificial insemination can be used when the man has a low sperm count 11. If the man is sterile a donor can supply sperm. 12. Intracytoplasmic sperm injection/ICSI can be used if sperm are defective/low in number 13. The head of the sperm is injected directly into the egg **(Any 10 – 1 mark each)**	10	CCCCCCCAAA	In full extended response questions, it is vital to use the language that is in the course specification – this is the key to scoring marks. Having a few extended response answers for each Area of study well-rehearsed is a good exam revision technique.

ANSWERS TO PRACTICE EXAM B

Question		Expected answer	Mark	Demand	Commentary with hints and tips
	B	1 Mother's blood pressure/blood type/blood tests/urine tests/general health check	10	CCCCCCCAAA	In full extended response questions, it is vital to use the language that is in the course specification – this is the key to scoring marks.
		2 Ultrasound (imaging/scan)			Having a few extended response answers for each Area of study well-rehearsed is a good exam revision technique.
		3 Dating scan/scan at 8–14 weeks is used to determine stage of pregnancy/due date			
		4 Anomaly scan/scan at 18–20 weeks for serious physical problems			
		5 Biochemical/chemical tests detect (physiological) changes of pregnancy			
		6 Marker chemicals/named chemical can indicate medical conditions			
		7 Marker chemicals/named chemical can give a false positive result			
		8 Diagnostic/further testing can follow from routine testing/named test			
		9 Amniocentesis/cells from amniotic fluid used to produce karyotype/to test for Down's syndrome/chromosome abnormalities			
		10 Chorionic villus sampling/CVS – cells from placenta/chorion used to produce karyotype/to test for Down's syndrome/chromosome abnormalities			
		11 CVS carried out earlier in pregnancy than amniocentesis			
		12 Allows immediate karyotyping			
		13 CVS has higher risk of miscarriage than amniocentesis			
		(Any 10 – 1 mark each)			

REVISION CALENDAR

You could use the Key Area calendar below to plan revision for your final exam. There are 24 Key Areas, so to cover two each week you would need a 12-week revision programme. Starting during your February holiday should give you time for the exam in May – just!

Key Area revised	Date	Questions completed (✓)	Green light for confidence! (✓)
Human cells			
1.1 Division and differentiation in human cells			
1.2 Structure and replication of DNA			
1.3 Gene expression			
1.4 Mutations			
1.5 Human genomics			
1.6 Metabolic pathways			
1.7 Cellular respiration			
1.8 Energy systems in muscle cells			
Physiology and health			
2.1 Gamete production and fertilisation			
2.2 Hormonal control of reproduction			
2.3 The biology of controlling fertility			
2.4 Antenatal and postnatal screening			
2.5 The structure and function of arteries, capillaries and veins			
2.6 The structure and function of the heart			
2.7 Pathology of cardiovascular disease (CVD)			
2.8 Blood glucose levels and obesity			
Neurobiology and immunology			
3.1 Divisions of the nervous system and neural pathways			
3.2 The cerebral cortex			
3.3 Memory			
3.4 Cells of the nervous system and neurotransmitters at synapses			
3.5 Non-specific body defences			
3.6 Specific cellular defences against pathogens			
3.7 Immunisation			
3.8 Clinical trials of vaccines and drugs			

Revision notes

Revision notes

Have you seen our full range of revision and exam practice resources?

ESSENTIAL SQA EXAM PRACTICE — National 5 | Higher

Practice questions and papers

- ✓ Dozens of questions covering every question type and topic
- ✓ Two practice papers that mirror the real SQA exams
- ✓ Advice for answering different question types and achieving better grades

NEED to KNOW — Higher

Quick-and-easy revision

- ✓ Bullet-pointed summaries of the essential content
- ✓ Quick exam tips on common mistakes and things to remember
- ✓ Short 'Do you know?' knowledge-check questions

How to Pass — National 5 | Higher

Scotland's most popular revision guides

- ✓ Comprehensive notes covering all the course content
- ✓ In-depth guidance on how to succeed in the exams and assignments
- ✓ Exam-style questions to test understanding of each topic

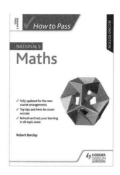

Our revision and exam practice resources are available across a whole range of subjects including the sciences, English, maths, business and social subjects.

Find out more and order online at www.hoddergibson.co.uk

HODDER GIBSON
LEARN MORE